THE UNION
SWITCH & SIGNAL CO.
Pittsburgh, Pa.

SAFETY ON THE RAILS:
The Union Switch & Signal Story

Joanne L. Harris

AnsaldoSTS
A Finmeccanica Company

Ansaldo STS USA, Inc.
1000 Technology Drive
Pittsburgh, PA 15219-3120
412-688-2400
www.ansaldo-sts.com

Joanne L. Harris, Author
Eric Fischer, Editor
Shannon Luchs, Research Assistant
Marg Watts, Research Assistant
Lou Anne Crosswhite, Proofreader
Russell Glorioso, Project Manager
Lauren E. Rockacy, Photography Manager
Greg Pytlik of Pytlik Design Associates, Inc., Designer

Table of Contents

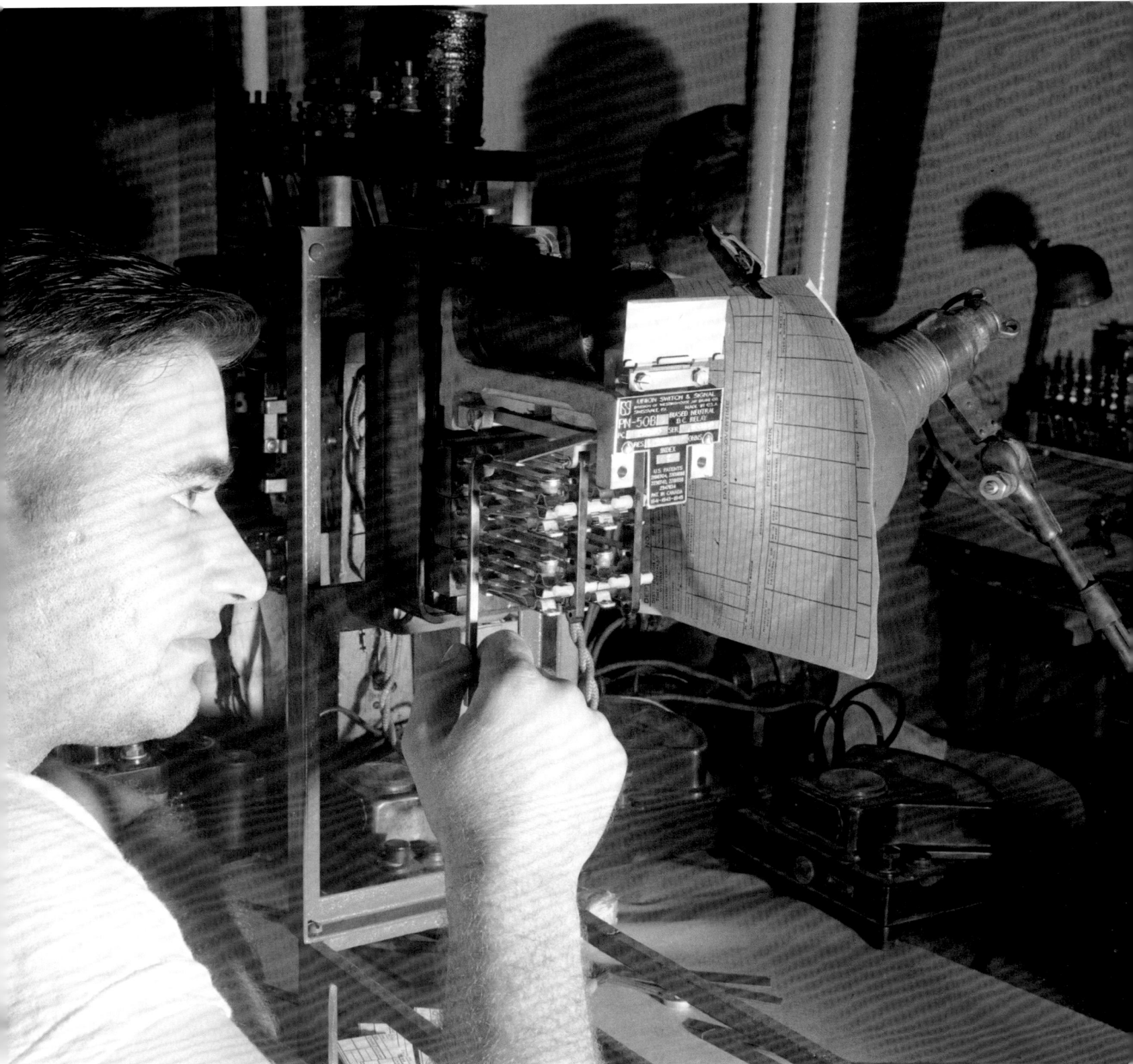

Foreword

The technological know-how embedded in the company is today, as in the past, in the legacy of its employees. A highly skilled group of professionals, knowledgeable of the rail signaling as discipline and as applied science. The company has driven innovation in signaling since its founding on April 13, 1881. Today, as in the past, the professionals at Union Switch & Signal (US&S) are engaged in many pioneering applications aimed at changing the industry's applications and operational needs, creating more industry firsts than we have in the past.

In commemorating the 130 years of continuous business activities, I thought it was important to document the great history of US&S and its employees. The book is based on the best available data researched for the purpose of this commemorative publication by the writer, to whom I extend my sincere appreciation, and a small support staff.

While great efforts have been dedicated to include every person, every significant project, and the many patented technologies that have enriched our history, we know that some have been missed. We regret this and we are certain in everyone's understanding. It has been almost impossible to reach all of the thousands of people who significantly contributed to the history and the success of the company.

The future of the organization is unquestionably good as it has its roots in such "technological greatness." The content of the book is proof of such an exuberant past and promising future.

This anniversary is a celebration to all the employees of US&S, now Ansaldo STS. I am grateful to all of you for the great efforts made and the success we have achieved together. I am particularly fortunate having had the opportunity to crown my professional career as the last serving President and CEO of US&S and to be part of such a distinguished roster of predecessors started by our founder, George Westinghouse.

I am grateful to all of the men and women of Union Switch & Signal / Ansaldo STS USA and compliment them for their outstanding accomplishments. This book is meant as a tribute to their work ethics, relentless approach to innovation and enduring dreams. A special thank you also to our Customers in recognizing that without them, our "dreams" would have not come true!

Alan E. Calegari, Ph.D.
President & CEO
Ansaldo STS USA, Inc.

Nomi delle minere in Furtelbach.
Pozzo di S. Guglielmo. Rumpapumo.
S. Gio. Furstenbauu. Ferrea porta. Cō=
uito degli ulmesi, S. Martino, Tre pax
zi uniti, Forno, S. Sangue, Martello,
culto de compagni.
Nella ualle Surbetz.
S. Michele, Verde Bosco, S. Giorgio
Copia, d' Argento.
Nella ualle prahegetz.
S. Filippo, S. Martino, Vite, Abiete uer
de, Armo monte, S. Vuillielmo.
Nella ualle uecchia Eckirch.
Beata uergine al freddo Fonte, S. Gia=
cobo.

Overleaf: An 11th Century medieval map of the silver mines in the region of Sainte Marie-aux-Mines (then Mancirch), Alsace, France, circa 1050.
© Hulton Archive/Getty Images

Insert Photo: An early mine truck that ran along a wooden railway in 1556, depicted in this illustration by Georg Bauer (aka Georgius Agricola) in his book, "De Re Metallica" ("On the Nature of Metals").
© National Railway Museum/Science & Society Archive Library

1   The Dawn of Rail Transportation

Throughout the history of mankind, man has relied on transportation to bring people and goods or services together. From the early transport of pack animals and animal-drawn wagons, to ships and modern day aircraft, the entire world has accepted the life-thread of transportation as the key to both survival and prosperity.

A farmer has to have a method to transport harvests, animals and supplies to and from his farm, or he will face spoilage and more importantly – failure. In the same way, as man developed opportunities to increase his prosperity by creating businesses and manufacturing products, access to the raw materials required to create the products and move finished goods was clearly essential to the very heart of the company's existence.

Some contend that the dawn of the railroad was established in its most rudimentary form by the Romans, who constructed wagonways of flat-topped stone blocks laid to support Roman chariot wheels. Others challenge that the Greeks' wooden beams, used to transport stone and marble for building their venerable structures, such as that of the Acropolis, were the birth of rail. Domesticated animals worked well for many centuries pulling wagons. In the 1300s, man devised the first wagonway – wooden beams laid down to traffic wagons filled with minerals or coal from mines, easing the burden of the animals that had to haul the loads over rough or muddy ground. The earliest known recording of such wagonways dates from about 1350 in a stain-glass window within the Cathedral in Freiburg, Germany. The unknown artist meticulously crafted a picture that depicts a narrow-gauge, silver mine track.

The First Railway Coach.

Resourceful mid-1600s men in Great Britain laid parallel tracks of timber for their horse-drawn carts.

Two centuries later, in 1550, artist Sebastian Muenster produced an illustration of a narrow-gauge mine railway at Leberthal, in the Alsace region of France.

It was not until the mid 1600s when several resourceful men in Great Britain thought to lay parallel tracks of timber that allowed horses to haul carts laden with heavy loads, such as coal, more expeditiously than they could over dirt roads. Here we truly find the beginnings of the modern day railroads. This practice continued into the early 1700s, when flat iron plates covered the wooden rails. Eventually lengths of angle iron replaced the plates, ensuring the wagons stayed "on track." Recognizing that the iron plates preserved the timber rails longer, the next logical progression was to form cast iron rims on the wheels in 1731.

The birth of the Industrial Revolution in the second half of the eighteenth century provided the economic background that afforded the true conception and development of the rail industry as we know it today. There would be, after all, no dramatic need of advanced transportation systems had it not been for the sudden surge in the creation of machine-based manufacturing and the introduction of steam power that this era provided. The Industrial Revolution found its beginnings in Great Britain, which at the time boasted the luxury of a wealthy middle class, a large surplus of capital and the readiness to spend it on new technology. Backed by financial encouragement from the Bank of England, British businessmen invested in machine-based manufacturing for the textile industry and developed advanced iron-making techniques. The steam power that drove the all-metal machine tools was primarily fueled by coal, such that the country was soon exploiting their extensive iron and coal deposits located in Northern Britain. This new demand for power, however, required improved transportation systems. Rivers and canals provided portions of the solution, but it was the expansion of railways that offered the most economical and efficient land transport.

Englishman William Jessop, a civil engineer born in Devonport, Devon, started his career designing canals and harbors across England and Ireland. He is deemed by most as the "Father of the Railway Track," having designed cast iron rails in 1789. At a mere three feet long, each length extended only from one sleeper to the next. These cast iron rails would soon inspire the invention of flanged wheels, designed to hold the carts to the rails; however, it was still authentic horse power that pulled the carts along the track.

The birth of the Industrial Revolution in the second half of the eighteenth century provided the economic background that afforded the true conception and development of the rail industry as we know it today.

Top: Richard Trevithick

© istock photo.com/Hulton Archive

Above: A shilling paid for a ride on Richard Trevithick's "Catch Me Who Can" steam locomotive in London in 1808.

A contemporary of Jessop, Richard Trevithick, also grew up in the Southwest of England. The son of a coal mining manager in Cornwall, he was surrounded by the industry and was fascinated by the technology of the steam-powered engines that were used to pump water out of the mines.

Trevithick built his first steam locomotive in December 1801 and carried the first passengers by carriage on "Captain Dick's Puffer" over a road surface in Cornwall. His engine design improved upon another inventor's steam technology – that of James Watts – by raising the steam pressure to a level that permitted him to use smaller cylinders and pistons. He patented the invention the following month.

Two years later, while working at the Penydarren Ironworks for Samuel Homfray, Trevithick tested a steam road carriage in the great city of London, reaching a maximum speed of eight or nine miles per hour. These achievements inspired his boss to take a bet against an unknown party that his employee's "Puffer" could not haul a load of ten tons of iron over a nine-and-three-quarter mile plate-way track. Trevithick responded by building his first steam rail locomotive, and on February 22, 1804, he added five wagons and 70 men to the wagered ten-ton load. He made the successful run from the Peny-darren Ironworks near Merthyr Tydfil, Wales, to the canal at Abercynon. The trek required multiple stops to clear trees and rocks from the path, but he achieved the trip nonetheless in four hours and five minutes, winning the bet for his employer.

Trevethick went on to design and build his renowned "Catch-me-who-can" steam locomotive, and ran it on a circular track in London. A financial failure due to lack of public interest in his shilling-per-head fee, he left the locomotive industry for good, spending the remainder of his life on one failed engineering project after another. He died in poverty in Dartford, Kent, but achieved the title the "Father of the Steam Locomotive."

In 1812, an inventor from Leeds, England, John Blenkinsop, came to believe that it would be impossible to haul heavy loads using smooth wheels on smooth rails. He devised a rail with cast "ears" upon which toothed wheels engaged to prevent slippage. This soon proved to be un-necessary for normal rail use, but it did, in fact, provide the technology for future step gradients in Switzerland that would require a rack-and-pinion rail. His efforts captured him the honor of having built the world's first steam locomotive for regular commercial use, with the capacity to pull a staggering 100 tons at three and a half miles per hour. The following year, another inventor named William Hedley dispensed with the tooth rack rail and built "Puffing Billy" and "Wylam Dilly," two locomotives used for hauling coal from the mines to the docks.

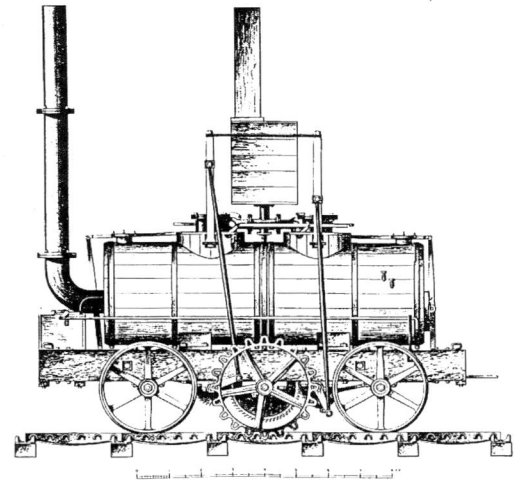

Inventor John Blenkinsop built the "Salamanca," a rack locomotive, in 1812 to haul coal from Middleton to Leeds in England.

Railroads came to America when the New Jersey legislature granted steamboat builder Colonel John Stevens III the first railroad charter in America in 1815 to build a railroad between the Delaware and Raritan Rivers. Stevens owned a large estate (roughly encompassing the current city of Hoboken, New Jersey), which at the time of purchase was described in the books as "William Bayard's farm at Hoebuck." Since he had a successful record as a steamboat designer, expectations were high for the first American steam locomotive. But when a year had passed and he had not built his railway, he sated critics by building his first steam locomotive, running it on a circular track on his estate – at a full 12 miles per hour.

The development of the steam locomotive was well underway back in England. George Stephenson, with a meager education, achieved a reputation at the age of 32 as an engineer. He saw a future in steam locomotives. Stephenson maintained his perseverance and curiosity in the realm of mechanical construction and won the respect of his superiors. With the financial patronage of Lord Ravensworth, owner of the colliery (coal mine) at Killingsworth, Stephenson built his first locomotive and named it "Blucher." During the lean economic times between 1814 and 1826, Stephenson was the only man in Britain who was building and advancing the technology of locomotives. His developments included the replacement of the noisy, unreliable gear drive with a sprocket chain. He later replaced the sprocket chain with coupling and connecting rods.

Stephenson was appointed as engineer to the Stockton and Darlington Railway Company in 1821, while the tugging between transportation modes (primarily canals versus road and rail) still brewed hotly.

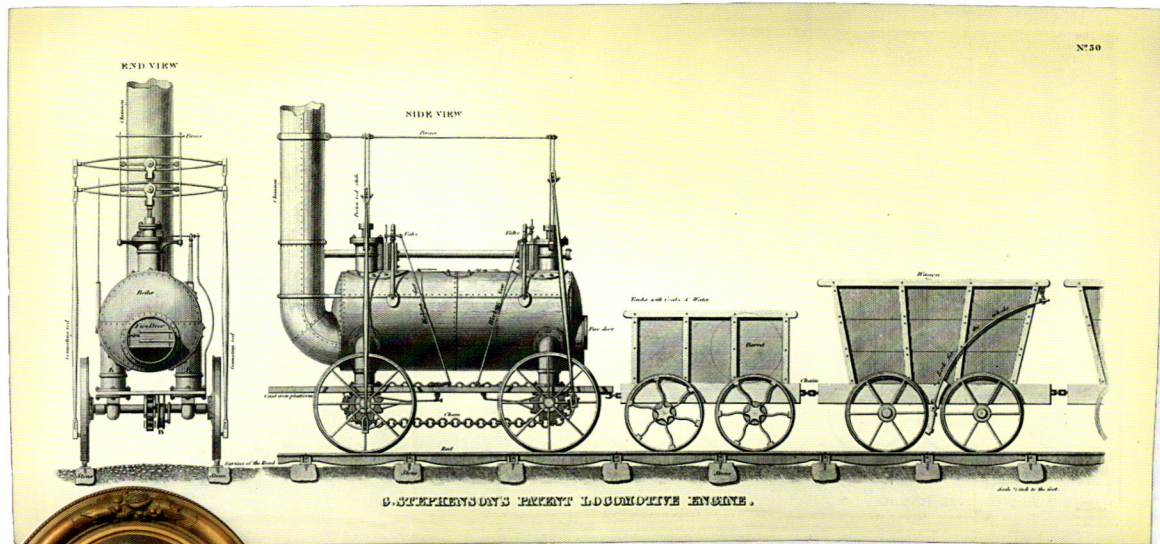

Rear and side views of George Stephenson's steam locomotive and railroad cars for the Stockton and Darlington Railway Company in the early 1820s.

George Stephenson
© istock photo.com/Hulton Archive

Stephenson's "Rocket" was the winning locomotive design for the Liverpool and Manchester Railway in 1826, netting him £500.
© istock photo.com/ivan-94

An Act of Parliament in 1821 authorized construction of the Stockton & Darlington line. The Act provided that anyone could put their wagons on the line and use their own horses, as long as they paid the Act's defined tolls. Stephenson tolerated the horse traction method until sufficient locomotives could be built. He laid out a 26-mile-long main line to connect the inland coal mines at Shildon to the Stockton docks in north-eastern England, creating the world's longest railway and the world's first passenger railway. Stephenson's next feat was to design the winning locomotive for the Liverpool and Manchester (L&M) Railway in 1826. "Rocket" won him 500 British pounds. It was soon apparent that the four-and-a half-ton engine was too small for the needs of the L&M, but it paved the way for new designs, resulting in an engine that was double the weight.

It was a foregone conclusion that the risks of traveling on a machine-driven locomotive far exceeded other transportation methods of the era. Thus, the L&M installed the first railroad signaling device, a wooden crossbar signal on a spindle, illuminated by a light on the track in addition to their flagmen. The board was turned parallel to the track when all systems were "go," and the light was white. If the locomotive engineer needed to "stop," the board returned to a 90-degree angle and the light was red.

Britain was not the only country thinking about the benefits of horse- and steam-traction railways. Ideas and innovations were happening all over the Western world. The Prussians were busy building Europe's first locomotives in 1816 and 1818 at the Royal Iron Foundry in Berlin, Germany. They used the Murray/Blenkinsop pattern, but neither engine was used for service outside of the railway yard. Austria and Hungary opened its first horse-traction railway in 1827 from Budweis to Trojanov, Czech Republic. Locomotives were not used until 1872. France opened its first steam-traction railway along a 38-mile line called the Lyon & St. Etienne. French engineer Marc Seguin built the line's first two engines, based on his visit to England to study and improve Stephenson's designs.

The Europeans continued to expand their steam rail systems in the 1830s with the Paris-St. Germain and two lines from Paris to Versailles. The first public steam railway in Germany ran from Nuremberg to Fürth, running with a British import Patenee 2.2.2 locomotive named "Der Adler" (The Eagle), complete with British driver. A mixture of horse traction railways and steam engine railways began to emerge in other countries, most of them initially built to serve the coal hauling industry, with passenger traffic not far behind. The Western European nations of Austria, Hungary, Ireland, Belgium, Italy and the Netherlands laid rail to move product and later passengers. Tasmania and Russia installed fledgling lengths of railroad in critical transportation regions.

Top: Peter Cooper

Above: The first steam locomotive on the railroad was the "Tom Thumb," built by Peter Cooper, making its debut in 1830.

The fragility of cast iron soon progressed to wrought iron in England, which proved to be a superior product against breakages. John Birkinshaw of the Bedlington Ironworks in England introduced 16-foot lengths of wrought iron rails in 1820, rails that could support a multi-ton steam locomotive. Across the proverbial "pond," Canada built what are considered the first North American cast iron rails of five-foot lengths for coal hauling in Nova Scotia in 1827 and in North Sydney in 1828. Both lines used horses. Canada's first steam railway did not appear for another eight years, when the 16-mile-long Champlain & St. Lawrence Railway line opened in 1836.

On America's Independence Day in 1828, Charles Carroll of Carrollton (so named to distinguish his signature on the Declaration of Independence from his father's name) laid the cornerstone of the Baltimore & Ohio (B&O) Railroad. The first 13 miles were opened between Baltimore and Ellicott's Mills, Maryland, for passenger and freight traffic on May 24, 1830. While horse-traction was used until mid-1834, the first steam locomotive on the railroad was Peter Cooper's "Tom Thumb," which ran on August 28, 1830.

Pennsylvania's Act of February 25, 1826, authorized the "commencement of the canal [the Pennsylvania Canal, connecting the Ohio and Delaware rivers] to be constructed at the expense of the state." Subsequent acts included the construction of the 36.69-mile-long Portage Railroad, which was opened in 1834. According to "*History of Pittsburgh*," the Portage Railroad carried 50,000 tons of freight

and 20,000 passengers in its first year. Passenger fare "from Pittsburgh to Philadelphia, by the Canal, Portage Railroad and the Columbia Railroad, was $12, and the trip consumed three days and nineteen hours."

When you look at the history of railroading in the United States and Great Britain, it is interesting to note that innovations in railroading, on tracks, trains and signaling equipment, had a continual progression facilitated by sending engineers back and forth between the two countries.

September 15, 1830, marked the date of the first death by a railroad train, when a carriage door of Stephenson's "Rocket" hit William Huskisson, a financier and Member of Parliament for Liverpool, as the Duke of Wellington opened the Liverpool and Manchester Railway. More railways meant more safety hazards, and by 1841, large disc and cross-board signals on up to 60-foot-high rotatable posts came into play as did the adaptation of the semaphore. Sir Charles Hutton Gregory first applied his semaphore invention to the New Cross, England, on the London and Croydon Railway. The "arms" of the semaphore were used for signaling in the same fashion as the arms of a flagman doing the same motions to "all clear," "caution," or "stop," and soon became the British standard.

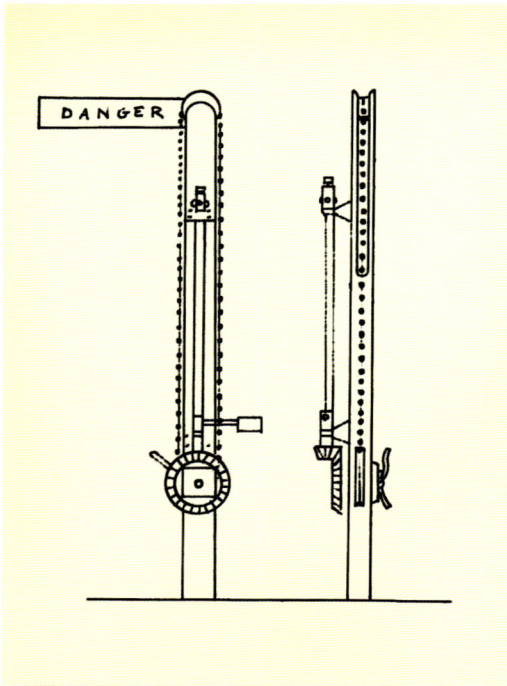

In 1841, Sir Charles Gregory introduced a semaphore signal with pulleys and wire rope that became the English railroad standard.

The first passenger train accident in U.S. railroad history happened on November 9, 1833 on the Camden & Amboy Railroad main line between Spotswood and Hightstown, New Jersey, when an axle broke on a car. One carriage overturned, killing two people and injuring 15, including Captain Cornelius Vanderbilt, formerly of the steamboat "New Brunswick." Captain Vanderbilt would later head more than a dozen railroads, including the New York Central. Ex-President John Quincy Adams remained uninjured in the next car ahead.

Railroad signaling was first used in the United States along the 17-mile-long New Castle and Frenchtown Railroad from New Castle, Delaware, to Frenchtown, Maryland, in 1832. They initially used a series of black or white flags on posts, which were soon replaced by a wooden basket covered with a white cloth. If the train was on time, the flagman ran the "proceed" basket up a 30-foot pole. A late or broken down train prompted the flagman to lower the basket, which signaled the engineer to stop. Large, hollow balls later replaced the baskets.

In 1841, George Stephenson suggested that the industry adopt a speed limit of 40 miles per hour, self-acting brakes and uniformity of signals on various lines across Great Britain. Edward A. Cowper designed a detonating fog signal that ignited a small quantity of gunpowder each time an engine's wheel ran over it. The single-needle telegraph, which used needle deflection to represent letters, presented a struggle for most stationmasters and signalmen in Britain, for the simple reason that many of them could not read or write. C. V. Walker of the South Eastern Railway devised a solution: a single-stroke bell telegraph system that used a simple stroke system: the number of bell strokes signified the various train positioning.

Inventors in Great Britain and the United States were hard at work to improve the safety of the railroads, and none too soon. Brazil, Switzerland, Hungary and Denmark (and a rail built in Denmark but annexed by Prussia), joined the ranks of rail building in the 1840s. The following decade Argentina, Spain, India, Austria,

Norway and Portugal joined the anticipated railroad path to profits.

Gloucester, England, resident Charles Wheatstone moved to London and partnered with William Fothergill Cooke to patent a telegraph to im-proved signaling, and by 1845, their invention was adopted on all English railway lines, becoming the foundation for the English block system. The

English traveler Alfred Bunn said of his American train experience, "When the matter of safety is consid-ered…America is lamentably below all other nations."

block system required "the establishment at spaced points of stations provided with traffic governing signals operated mechani-cally by block operators in accordance with the schedule of operations fixed by the timetable and with information communi-cated to the operators from adjacent stations by electric telegraph. The system "provided protection against collisions from following trains, but relied upon the memory of the operators as to the state of traffic in the block system."

America lagged behind on many safety issues. According to an excerpt in the book "Death Rode the Rails," English traveler Alfred Bunn said of his American train experience, "When the matter of safety is considered… America is lamentably below all other nations." When another English traveler, Charles Weld, "complained that reckless running had caused the train he was on to derail, he found that most passengers applauded the crew's effort to make up time."

As some passenger trains in the United States were now running at 30 miles per hour (44 feet per second), Massachusetts, New Hampshire, Connecticut and New York instated requirements for the construction and maintenance of road crossings, fencing and line inspection. Having requirements and sustaining them were two different stories. In 1852, New York's line became so abominable that the *Ogdensburgh Republican* lambasted them in print, stating, "A proposal has been made to our State Legislature to abolish hanging and substitute for punishment a ride upon some of our railroads."

Notwithstanding, England's Edward Tyer made advances in the block telegraph system that developed into the new position of the train dispatcher. Sending train orders via telegraph allowed dispatchers to modify train schedules, reroute or stop trains, or report delays. Soon afterwards, Tyer went on to invent the step-by-step alphabetical telegraph for communicating between signal boxes. The new technology was the precursor to train describers.

Four and a half million immigrants were now living in America, yet as of 1850, there were still no tracks laid west of the Mississippi River. But not all cities east of the Mississippi were prospering in the renewed industrial revolution. Pittsburgh, Pennsylvania, is one example of an East Coast city that was in terrible debt. According to "*History of Pittsburgh*," the city's bonded debt was over a million dollars in 1850, after paying

Top: Charles Wheatstone

Above: The four-needle tele-graph, circa 1837, provided Cooke and Wheatstone the baseline studies to develop the five-needle. The later version was recognized as the first functional electrical telegraph.

Photo credit: BT Heritage/UK

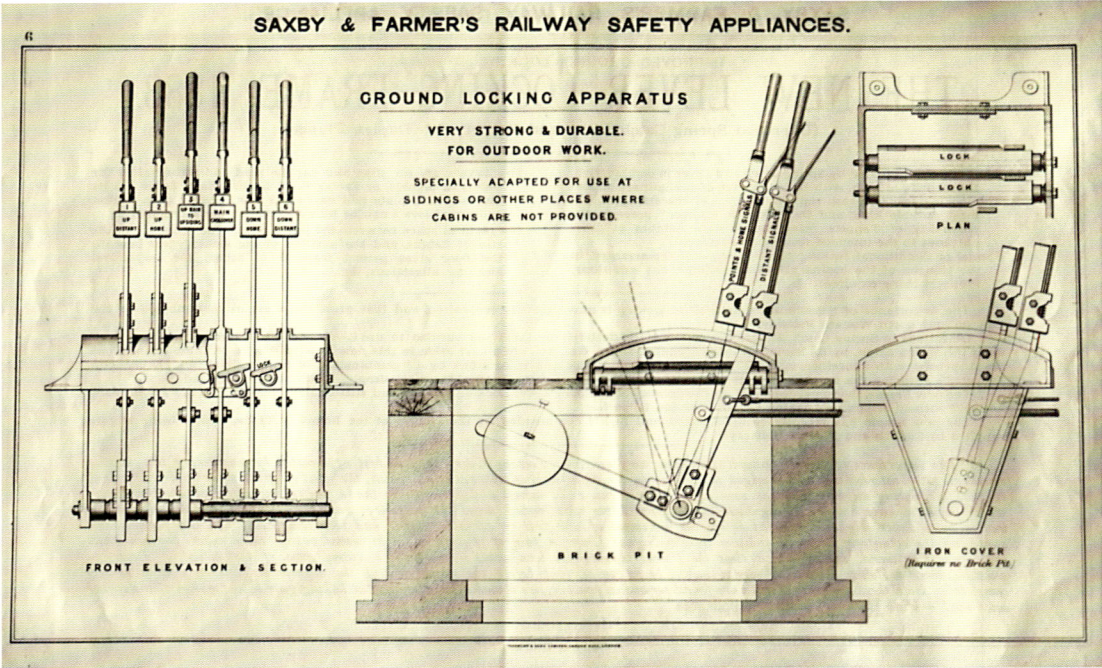

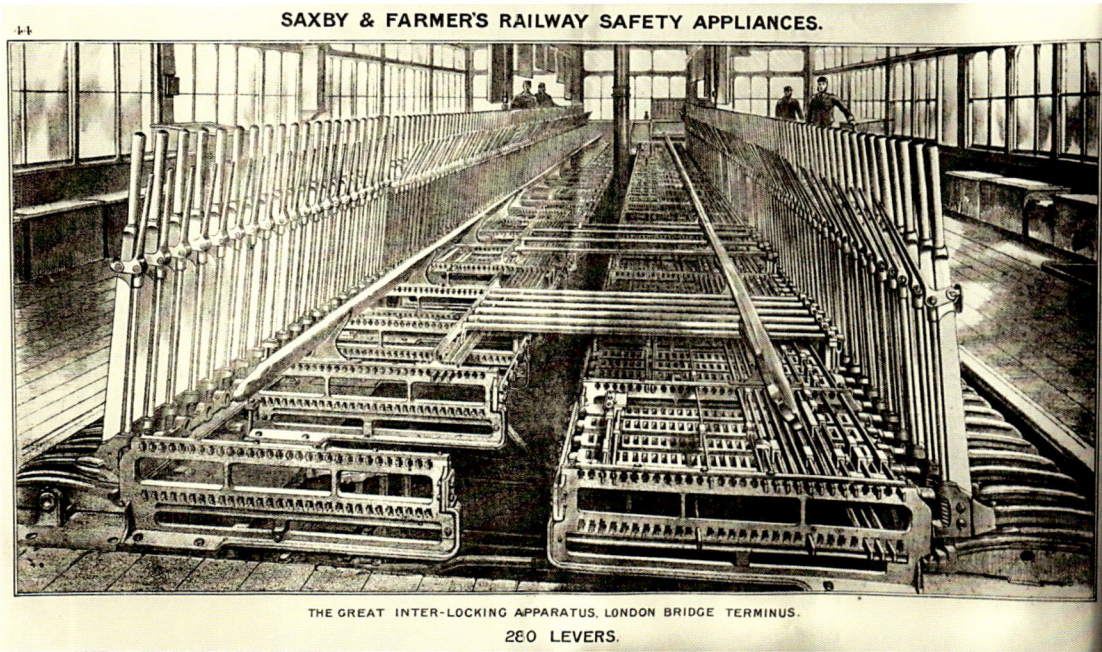

The 1888 Saxby & Farmer catalog featured these intricate interlockings that quickly grew in popularity throughout England, Europe, Russia and North America.

for county improvements and for "city improvements which should have been borne by individuals, business corporations and firms." They took a complete step off the ledge when an Act was passed that allowed the city to lend itself further funds by issuing $1,800,000 in bonds to finance five railroads: the Ohio & Pennsylvania; Pittsburgh & Steubenville; Allegheny Valley; Pittsburgh & Connellsville and the Chartiers Valley Railroad. Now nearly $3,000,000 in debt, outside businesses and capital investors would not touch the city until the railroads were complete and showed hopes of local business revival.

Interlocking is... "an arrangement of switch, lock and signal appliances so interconnected that their movements must succeed each other in a pre-determined order."

But business did continue, as did the development of railway safety equipment. While early cars might accommodate up to 20 passengers, by 1850 they could serve 60 travelers. This created higher profitability for the railroad, but also increased the risk of disaster, as a derailment or collision often damaged only one car (evidenced, for example, in the wreck of 1833). Such increased risks inspired further investigation into signaling systems. The first interlocking systems appeared when Sir Charles Hutton Gregory invented the basics of interlocking, but his device did not completely interlock, as the switchman used his feet to work the signals and his hands to control the switches. The word "interlocking," after all, is defined as "an arrangement of switch, lock and signal appliances so interconnected that their movements must succeed each other in a pre-determined order."

The British development of interlocking systems snowballed from there. Austin Chambers, engineer of the North London Railway, created the first true interlocking in 1859; the firm of Stevens and Son produced the first interlocking system that locked distant signals the same year. Soon after, former blacksmith John Stevens and his son founded the world's first signal manufacturing company, the Darlington Works. John Saxby invented a mechanical interlocking system that controlled signals and switches in a single operation from a central point. He received his first patent in this field in 1856, when Great Britain had about 9,000 miles of rails and Egypt opened her first line. Seven years later, he partnered with John Stinson Farmer, and by 1872, the Saxby & Farmer factory near London employed some 700 skilled artisans and others, with foreign customers in Italy, Spain, Portugal and Russia.

Interlocking was now a reality, and the technology was soon adapted, improved upon and manufactured in America.

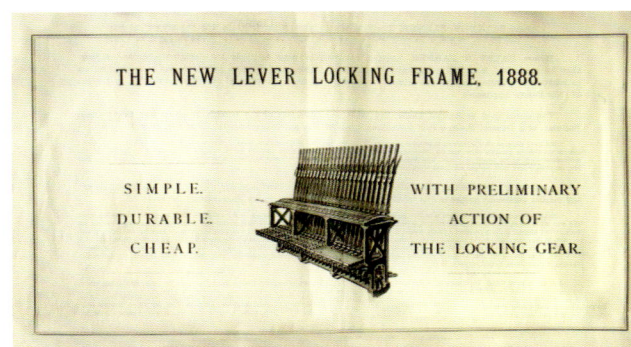

THE NEW LEVER LOCKING FRAME, 1888.

SIMPLE.
DURABLE.
CHEAP.

WITH PRELIMINARY
ACTION OF
THE LOCKING GEAR.

This 172-lever interlocking was installed in 1920 at Dolton Tower in Dolton, Illinois. The fact that it is still in service today gives credence to the durability of Union Switch & Signal equipment.

Photo credit: © Jon R. Roma

2 American Pioneers of Automatic Signaling and Interlocking

The 1870s emerged with the big railroad boom between 1866 and 1873. Government land grants totaling over 170,000,000 acres, along with generous subsidies, were made available to the railroad companies by 1871, and 56,000 miles of track were laid across the country. The rail industry was at the time the largest employer in the country, with the exception of agriculture.

The railroads in the United States began to carry the U.S. mail in 1831 and by 1838, the first U.S. Railway Post Office was put in service. By 1834, Postmaster General William T. Barry contracted several trains to carry the "Great Eastern Mail." Railroads were spreading like wildfire across the United States and Canada, albeit neither country would have a complete continental connection until 1869 and 1885, respectively. On May 10, 1869, the Union Pacific and the Central Pacific met and drove the last spike (a golden spike, at that) into a California laurel tie, marking the end of the valiant Pony Express and the Overland mail stages services; their era would soon be a romanticized memory in history.

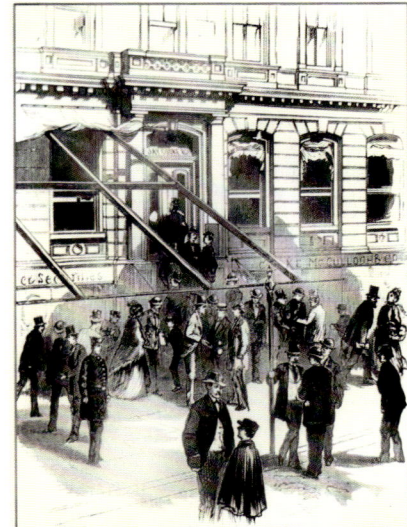

Jay Cooke & Company, a Philadelphia bank that held a large interest in bonds for railroads, was the first bank to collapse in the Panic of 1873.

The 1870s, however, soon faced the onset of an economic turmoil worldwide. Overexpansion – including the rail industry – in Germany and Austria sent their countries into a depression. England's financial standing sank to a drastic low, and the United States did not escape the monetary crisis. The country had survived the Black Friday panic of 1869 and the Chicago fire in 1871. It was in 1873 when the German empire stopped coining silver "Thalers" that the silver mines were severely impacted in the United States. The price of silver plunged, and to make matters worse, the United States government passed the Coinage Act of 1873, stating that it would no longer back its currency with silver. Hence, with a gold-only standard in place, rising interest rates and an aversion to investments in long-term bonds quickly brought banks to their knees. Beginning with Jay Cooke & Company, a Philadelphia banker and financier for the Northern Pacific Railway, and Livermore, Clews and Company banks, which was heavily in the bond business, one bank after another failed. Even the New York Stock Exchange closed its doors for ten days.

Early English railroads were required to employ a flagman to stop every train at a crossing, but that requirement might be waived if the crossing was interlocked.

Eighty-nine of the 364 U. S. railroads went bankrupt, and construction on the Northern Pacific Railway came to a screeching halt. It would be a lean six years, with 14 percent of the country out of work due to failing businesses and a lack of construction. Yet the railroads that survived continued to build.

In the midst of all the anxiety, state governments began to take railroad safety issues more seriously and established regulations that would continue to develop when initial practices did not suffice. Massachusetts established a Board of Railroad Commissioners in 1869 to investigate accidents and to make recommendations, but as it had no other power, little progress was actually made. Michigan began to regulate crossings of railroads by other lines, and other states soon followed. The law required trains to stop at crossings and that rail companies had to keep a flagman at each crossing; if the crossing was interlocked, the requirement might be waived.

America's first manual block system used banner box signals to safely guide their trains.

American railroad companies began meeting to discuss issues that affected rail business. A collective of General Managers and ranking railroad operating officials held their inaugural meeting in 1872. Originally known as Time Table Conventions, the group changed the meeting name in 1875 to the General Time Convention and in October 1892, became known as the American Railway Association (ARA). In 1919, ten separate groups of operating officers merged with the association and executed their tasks as divisions or committees of the association. On October 12, 1934, the ARA joined with several other railroad industry trade groups to form the Association of American Railroads (AAR).

In Britain, high traffic and cheap wages for telegraphers encouraged the use of manual or controlled-manual block systems. Between 1873 and 1880, the total amount of blocked rail miles soared from nearly 40 percent to 80 percent. The following year, British Parliament made blocking mandatory. The dramatic decline in passenger deaths from collisions that followed the Act was proof that block systems were indeed a viable part of the answer.

Railroads were still primarily single track, and safety companies developed the beginnings of "read and repeat" train orders with special forms. An agent would deliver one copy of the order to the conductor and another to the engineman and often required them to read the order back to the agent. This system increased productivity for railroads, reducing delays and avoiding collisions.

This new decade brought a plethora of advancements in interlocking machines, a rise in head-on collisions, and mounting regulation in the United States. American civil engineer Ashbel Welch, who had introduced the block system to the United States railroads in 1865, also introduced interlocking to America after he investigated the British technology in England. He proposed the system to the United New Jersey Canal and Railroad Companies, of which he was the General President and Chief Engineer. He was authorized to test a Saxby & Farmer machine No. 905 interlocking system on a line that was leased to the Pennsylvania Railroad (PRR) at the Trenton, New Jersey Station in 1870. The plant was dismantled some three or four years later due to expiration of the lease and the small amount of traffic at the junction. He installed a second Saxby & Farmer machine in 1875 at East Newark, New Jersey, on the PRR.

> "It is a trite axiom, that two bodies cannot occupy the same space at the same time. The duty of the railway signalling engineer may be said to endeavor to prevent two bodies, which are moving at high velocities, from seeking to violate this law of nature."
>
> — Richard C. Rapier
> in "On the Fixed Signals
> of Railways," 1874

The Railroad Gazette began to record the number of collisions after 1873, and reported that one of the deadliest years in the 19th century history of rail was 1877 with a record 207 passenger fatalities. (This year was perhaps surpassed only by the year 1853, during which more than one hundred major rail accidents took the lives of 234 people and seriously injured 496 more, including a Norwich, Connecticut, disaster that killed 45 people and a wreck in Illinois that took the lives of 18 people.) The causes included a mixture of neglect, obsolete safety equipment – including the Loughridge straight air brake – as well as crossings and yard work collisions.

In one of the 1877 disasters, a Baltimore & Ohio (B&O) freight train left Garrett, Indiana, heading east toward Republic, Ohio. The engine broke down on the single track just west of the station and the westbound express was due at the station in 10 minutes. The conductor failed to send out flag protection for his train, and the express plowed into the freight, killing 13 people. The B&O was still using the straight air brake that had been obsolete since 1872, and was slower than the Westinghouse automatic version. The engineman of the express had seen the freight about 1,000 feet away while he was running about 43 miles per hour; an automatic brake should have stopped the train. The accident also revealed the need for a "proper system of clearly visible fixed signals in conjunction with the block system," reported the Railroad Gazette.

Six years later, Arthur Wellington of the Engineering News pointed out that only 11 passengers had been killed in all forms of British train accidents and passenger risks had fallen by two-thirds in 12 years. He calculated that American lines were at least five times as dangerous, and he also called on carriers to begin "blocking."

Thus began the search for safer systems. The leading American pioneers in automatic signaling systems were Thomas S. Hall, Franklin Leonard Pope & Stephen C. Hendrickson, William Robinson, David Rousseau and Henry Spang. Each had much to offer the industry, but many times they found themselves in the throes of yet another patent lawsuit. It stands to reason that even beyond the normal scope of patent protection, that these companies – and those to follow – closely guarded their territory and inventions as the growing number of suppliers quickly outpaced the number of potential rail customers.

Thomas S. Hall developed a wire circuit system that was installed on 16 miles of the Eastern Railroad and on the New York and Harlem Railroad in 1871. His system used enclosed disc signals that housed a banner that moved in and out of an aperture within the housing. Two treadles, or circuit "breakers," were mounted along the track at each end of the block for actuation by the train wheels. When a train entered the block it tripped the first treadle, breaking the circuit and setting the signal to "stop." When the train exited the

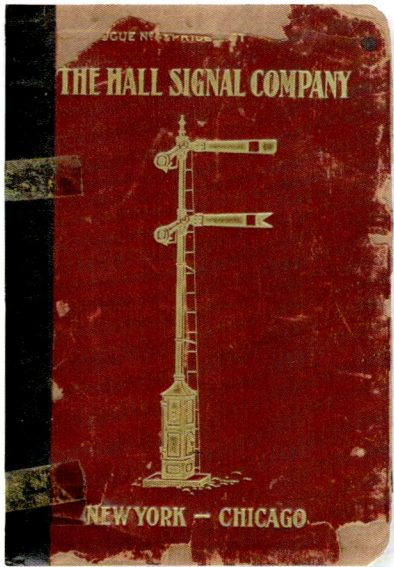

Hall Signal Company 1901 catalog

Photo credit: © Zachary Gillihan Collection

block, it tripped the second treadle, closing the circuit and setting the signal to "clear." The danger of this open circuit system
was that it was not failsafe; if a train broke in two, the signal
would automatically clear when the first section left the block.

Hall's early patents were assigned to Hall's Electric Rail-
way-Switch and Drawbridge-Signal Company of New Haven,
Connecticut. The company was reorganized in 1873 under the
name Hall Railroad Signal Company with a Connecticut charter.
The Electric Railroad Signal Company sued the Hall Railroad
Signal Company in 1875 on an infringement of Franklin L. Pope's
patent 140,536, involving an installation that used a central
battery connected to multiple signals. In 1885, the Hall Company
was found not guilty of infringement. After Thomas S. Hall's
death in 1880, the Hall Railroad Signal Company united with the
Wharton Railway Switch Company of Philadelphia, which grew
stagnant until W. P. Hall and A. W. Hall, two of Thomas Hall's sons,
reorganized the company in 1889 as the Hall Signal Company,
chartered in Maine.

Franklin L. Pope and Stephen C. Hendrickson joined forces
to develop an automatic system using normally-open track circuits and signals operated by clockwork that was controlled by
electric magnets. Pope was one of the first practicing electrical
engineers in the United States, and one of the country's first
patent attorneys. He had previously partnered with Thomas
A. Edison to form Pope, Edison & Company, Electrical Engineers
in 1869, and during the short existence of the company (less
than two years), they invented a one-wire printing telegraph.
Pope and Hendrickson were employed by the Pennsylvania
Railroad in 1872. Pope devised a system in which one battery
operated both primary and secondary clockwork signals.
Hendrickson designed the mechanical features of this system, which was electrically controlled, but
mechanically operated. Their patents were assigned to the Electric Railroad Signal Company. After the
aforementioned lawsuit against the Hall Railroad Signal Company, the patent and a number of other patents
developed by one or both of the engineers, was assigned to The Union Electric Signal Company prior to
1881. The courts dismissed the action on the grounds that Hall was the prior inventor of the patented system.
After the Union Electric Company acquired the Pope & Hendrickson Patent, they elected Pope as a Director
of the company, where he remained active for many years. He later became one of the incorporators of the
Westinghouse Electric Company, specializing in patent matters.

In later years, Pope became editor of the "Electrical Engineer" magazine, and two years later he
became the second president of the American Institute for Electrical Engineers (AIEE). Ironically, this man,
who spent his life's work in the field of electricity, was electrocuted by a 2,100-volt overhead distribution
line in his New Jersey basement in 1895.

One creative farmer requested payment
for a pig that found its way onto the tracks:

My razorback strolled down your track
A week ago today;
Your 29 came down the line
And snuffed his life away.
You can't blame me – the hog, you see,
Slipped through a cattle gate,
So kindly pen a check for ten
The debt to liquidate.

To which the claim agent replied:

When farmer's swine get on the line
Where trains have right of way,
And when a stake he tries to make
By clamoring for pay.
He wastes his time in penning rhyme;
The claim's not worth a fig.
It seems to me—and you'll agree—
He should have penned the pig.

Opposite: Thomas Hall devised banjo-style signals, like this one on the Reading Railroad.
© Jon R. Roma Collection

**Dr. William Robinson,
Father of Automatic block
Signaling.**

Irishman William Robinson graduated from Wesleyan University in Middletown, Connecticut, in 1865. He worked in the educational system for a short time before delving simultaneously into the Pennsylvania oil business and the development of an automatic signaling system for railroads. Robinson first devised a wire system using track instruments. This invention, as he detailed in his October 25, 1870, patent, "relates to electro-magnetic apparatus for signaling by the train, while in motion, breaks the main track on a line of railroad, as by the displacement of a switch or opening of a drawbridge, or through the operation of adjuncts which are employed in connection with these devices, to effect change in their position." The invention also included an "irregular spacing of the circuit-breaker or breakers, for operation in connection either with an irregularly spaced or ordinary circuit-closer; to be indicated by the intermissions in the sounding of the alarm the direction in which the train is travelling."

Robinson built a detailed model of this system and exhibited it at the 1870 American Institute Fair in New York City. He sent out leftover advertising pamphlets from the Fair to a number of railroad companies. This first known instance of automatic signaling advertising earned him a meeting with William Ashbridge Baldwin, then general superintendent of the Philadelphia and Erie Railroad in Pennsylvania. He subsequently secured an opportunity to install a full-scale, trial system on the Philadelphia and Erie Railroad at Kinzua, Pennsylvania, the same year.

The open circuit system soon revealed several inherent design defects. The forward section of a uncoupled train would trip the circuit, showing an "all clear" when the remainder of the train still monopolized the track. A train could enter a section from the opposite end or from a siding, blocking the track, while not affecting the signal. As well, broken line wires, failed batteries or a broken connection would prompt a potentially dangerous "all clear" signal.

William Robinson then experimented with a closed-rail, circuit system patent that he exhibited in a model format at the Erie, Pennsylvania, State Fair in 1872. The new failsafe track circuit system relied on every car and pair of train wheels to control the signal throughout the block; the electric current kept it in a "clear" position. If the power was lost and there was no electricity on the semaphore, gravity would literally pull the semaphore signal down to the "red" or "stop" position. Today, the rail industry would say, "It works on a 'failover to red' basis." The Kinzua installation changed over to the new system and added an additional installation at Irvineton, Pennsylvania, on the same line.

Robinson made other test installations before moving to Boston for a trial installation on the Boston & Lowell Railroad and several further regional installations. He installed a variant of the system in California's Tehuantepac tunnel in 1877, the first instance where two separate rail circuits were used in a cascade fashion to control a signal. Today, the boundary between these circuits is known as a "cut section." The initial installations comprised a magnetic signal mechanism, housed in a small building alongside the track, and a signal banner mounted on the roof of the building. Robinson's first revenue-service track circuit on 10 miles of the Fitchburg Railroad in Northern Massachusetts utilized Oscar Gassett's magnetically controlled, mechanically operated clockwork signals.

Early electric semaphores appeared around 1897, soon replacing banjo-style and clock-work disc signals.

The installation was made under a contract entered into with the Fitchburg Railroad on behalf of a proposed company, which shortly after formation assumed the Fitchburg Railroad contract. In late 1878, Robinson organized the Union Electric Signal Company, of which he was the president and owner. He assigned nine United States Patents, including his basic closed-circuit patent 130,661 and other related signal devices to this company and subscribed to 98.3 percent of its stock. He then spent 15 months traveling in Europe, Egypt and Palestine. Robinson was elected a director of the company at its formation, but in October 1879 C. H. Jackson replaced him. After that time Robinson was never active in the management of the company nor did he participate in the future developments of railway signaling.

"Perhaps no single invention in the history of the development of railway transportation has contributed more toward safety and dispatch in that field than the track circuit. By this invention, simple in itself, the foundation was obtained for the development of practically every one of the intricate systems of railway block signaling in use today, wherein the train is, under all conditions, continuously active in maintaining its own protection.

In other words, the track circuit is today the only medium recognized as fundamentally safe by experts in railway signaling whereby a train or any part thereof may retain continuous and direct control of a block signal while occupying any portion of the track guarded by the signal."

– Third Annual Report of the Block Signal and Train Control Board to the Interstate Commerce Commission, dated November 22, 1910.

The Robinson closed-rail circuit of 1872 and 1874 (under a re-issued patent) forms the basis of every modern-day automatic electric, electro-pneumatic and electrically controlled fluid pressure system in the world.

By 1890, rail lines would rely on semaphores during the day and lights at night. When a train entered the block, it would short the circuit and shift the semaphore to a horizontal position, stopping following trains. It would also shift the distant signal on the previous block to "caution," giving any following train time to stop. If the semaphore's distant signal displayed "caution," the engineman knew that the next block was occupied and would decrease his train speed in preparation for a stop. This system was failsafe, for anything that broke the circuit – a failed battery, equipment left on the track rail placed by train wreckers or a broken rail – would cause a counterweight to automatically shift the semaphore to "stop."

Manual block signals were nearly universal in Britain by 1880, but American carriers mainly stayed with the train order system. Control with train orders worked best with light traffic and an experienced, well-disciplined labor force. But a labor market characterized by rapid growth, high turnover and agency problems interfaced with a rising traffic density, and it was noted that the train order system was now placing too much responsibility on dispatchers, telegraphers and train crews.

In 1906 Robinson published a booklet entitled "History of Automatic Electric and Electrically Controlled Fluid Pressure Signal Systems for Railroads" and in 1907 he earned his Doctorate at Boston University. Dr. Robinson died January 2, 1921. The Signal Section of the AAR published a booklet entitled "The Invention of the Track Circuit" in 1922, as a memorial to William Robinson, in which the prominent points of his booklet were reproduced. Robinson was closely associated with Franklin L. Pope, Oscar Gassett and Israel Fisher during his Boston years. In spite of their relationship, he only gave Pope credit for bringing out an open circuit system, Gassett for his success as a promoter and Fisher as an excellent mechanic who was instrumental in making a large part of the signal apparatus which Robinson installed on New England roads between 1876 and 1878.

David Rousseau developed and patented an automatic block system that was another form of treadle or circuit-closing signal device. A circuit-closer located underneath the rail operated and controlled the signal. The disc was set at a right angle to the track to indicate "stop;" when set parallel to the track (thus not visible), it indicated "proceed." At night a lamp was placed in the back of the case, glowing red for "stop" or white for "proceed." A revolving vertical shaft controlled by a magnet and its armature operated the disc. A wind-up counterweight protected the disc from remaining at "proceed" in case the current was interrupted.

Rousseau provided block indicators at the stations that repeated the position's block signals. His system was installed on part of the temporary tracks laid down for the Philadelphia Centennial of 1876, but operators manually controlled the signals in signal cabins through the use of push buttons. The automatic version of the system was installed on the New York Central. By 1879, it was reported to have been in successful operation for nearly four years. The system provided two danger signals, one a distant signal some 1,000 feet before the signal at the section to be entered. The Union Switch & Signal Company acquired eleven Rousseau patents in August, 1881.

"The rear end collision at Revere Station on the Eastern Railroad (now a part of the Boston and Maine) upon the evening of August 26, 1871, between an express and accommodation passenger trains, killing twenty-nine and injuring fifty-seven, greatly stimulated invention and brought forth several automatic electric block systems." — Henry W. Spang

Henry W. Spang began experimenting with automatic railway signaling from 1872 until 1875. He developed a wire system operated by the tread of the wheels in conjunction with an insulated iron bar laid parallel to one of the track rails. A battery connected to ground at one side was connected through the signal's electromagnet to the iron bar (or auxiliary rail) that was insulated by the ground. The adjacent track rail was connected to ground, and a wheel tread bridged the space between the rail and the parallel iron bar to close the circuit whenever a train passed over the bar. Sprang tested his first automatic block system in June 1872 on the Lebanon Valley branch of the Philadelphia and Reading Railroad.

Drawings of David Rousseau's patented circuit closer of 1873 were a component of his automatic block system, first installed on a section of temporary track leading into the 1876 Philadelphia Centennial.

According to Spang, wire systems became obsolete, "principally for the reason that safety signals were given by induced electricity during thunderstorms, crosses with telegraph wires and also when a detached portion of a train was on a block section." In March 1881, the American Railway Signal Company was formed to manufacture and introduce his signaling system; however, his eight patents were assigned to The Union Switch & Signal Company on June 8, 1881. In 1902, Spang wrote a 75-page booklet called "A Treatise on Perfect Railway Signaling," which reviewed the history of track circuit signaling and concluding with an argument in favor of the "normal-danger" block signal system as being the "perfect" system.

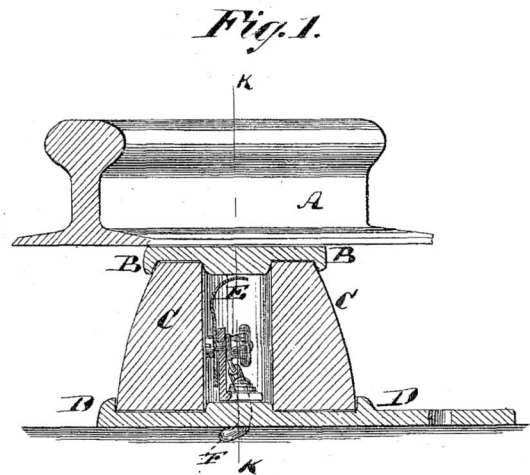

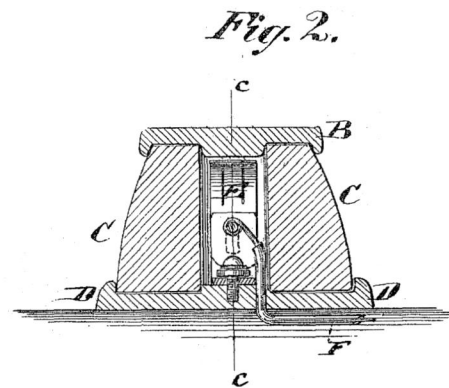

The great railroad riots of 1877

The year was 1877. Forty B&O Railroad men in Baltimore joined striking firemen and brakemen on July 16, who had refused a ten percent wage reduction. The train crews were replaced, but a rabble of troublemakers joined in and stopped all trains three miles from town. Before long, the trouble spread like wildfire, first to Martinsburg, West Virginia, and then Pittsburgh. When the mobs ignored local and state authorities, U.S. President Hayes ordered rioters to disperse. He dispatched 250 soldiers to Martinsburg.

Troops were sent to Martinsburg, West Virginia, after strikers blockaded locomotives on July 16, 1877. The riots quickly spread from Baltimore, Maryland, to Pittsburgh, Pennsylvania in 1877.
Engraving from Harper's Weekly

The mobs had been incensed with the railroad companies for years, provoked by news articles that the "Railroad Vultures" were "constantly preying upon the wealth and resources of the country." Soon, the B&O, the Pennsylvania Central, the New York Central & Hudson River Railroad, the Lake Shore & Michigan Central, and the Erie Railroad were all affected. Ironically, the strikers were soon lost in their own battle.

By the next morning roughly 2000 cars were tied up in Pittsburgh's downtown freight yard, and thousands of livestock dying in cars at the East Liberty Stock Yards. Philadelphia troops arrived on the 21st and Sheriff Fife, Superintendent Pitcairn by his side, read the Riot Act, with warrants to arrest 15 of the ringleaders. The strikers hurled stones and as a revolver shot fired into the ranks, General Brinton ordered his men to fire, killing about 20 people, including several children and onlookers.

That night, several thousand men set fire to oil cars and a car filled with coke, shoving them up against the roundhouse. The building soon ignited. The troops were finally able to emerge from the roundhouse, and forged through the city to get to the U.S. Arsenal for safety, but the Commandant turned them away.

The mob broke into liquor stores, ending the conflict as troops found them sleeping off the resulting drunkenness. The roundhouse — and over 100 locomotives beyond Union Depot — were destroyed. Machine shops and railroad offices were burned to the ground. The flames licked their way toward the Union Depot on Sunday afternoon. Rioters broke into the Depot Master's office, using books and papers as tinder to set the office on fire. These were no longer a handful of unhappy crewmen, but a mob intoxicated with anger as well as liquor.

On Monday morning, the City of Pittsburgh formed a Committee of Public Safety, who held a public address and published a decree to restore order. They urged businesses to keep their employees at work and onlookers to stay off the streets, and formed a Vigilance Committee for citizens who wanted to protect the city. According to The History of Pittsburgh, the decree stated that "[the riot] gave demagogues and bad men the opportunity to play upon the passions of the masses, and what was a mere, in one sense, harmless strike of a few dissatisfied railroad employees, who intended no violence, became the terrible riot… Claims of $4,100,000 were made against Allegheny County for damages, and commissioners settled for $2,772,349.52. Of this, $1,600,000 went to the Pennsylvania Railroad. "In addition to the buildings [burned], 1,383 freight cars, 104 locomotives and 66 passenger coaches [were] destroyed. Twenty-five persons in all were killed."

Impatient rioters broke into the Depot Master's office, urging the fire that finally consumed Pittsburgh's Union Depot.

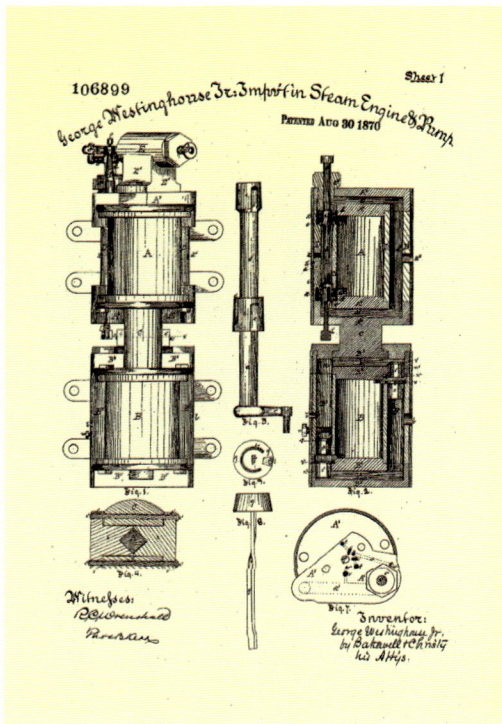

George Westinghouse sold his patent for this steam-driven air pump to Union Switch & Signal in 1891. The pump operated track switches and banner signals, but the system was somewhat inefficient as the air admitted at the machine had to charge the entire line to the operation device as well as the operating cylinder for the device.

The pioneers in the field of interlocking in the United States were John M. Toucey and William Buchanan, who founded the Toucey and Buchanan Company, David A. Burr, and C. H. Jackson. Toucey and Buchanan were the General Superintendent and Superintendent of Machinery, respectively, of the New York Central & Hudson River Railroad. In 1874, they designed and installed the first American-built interlocking plant at Spuyten Duyvil (Spitting Devil) Junction on the Hudson River Railroad just north of New York City. The machine had a two-notch plate on a hinge paralleling the motion of the lever, instead of a squeeze-handle latch. A foot treadle plunger lifted the plate. They installed the same design on the lines leading into the Grand Central Station in New York in 1875, on the temporary tracks (three plants) of the 1876 Centennial and on the Metropolitan Elevated road in New York City. At the 1876 Centennial at Philadelphia, Saxby & Farmer and the London, England, firm of John Brierly & Sons displayed their two interlocking machines. The Toucey & Buchanan Company later acquired the American rights of these two English firms, renaming the company The Toucey & Buchanan Interlocking Switch and Signal Company in May 1877, and assigning their patents to the company. The Interlocking Switch and Signal Company took over these patents just prior to the consolidation of that company with The Union Electric Signal Company.

David A. Burr developed and patented an interlocking system for operating switches and signals, and indicating their positions. His system was installed and placed in service in July of 1876 on the temporary tracks (Mantua Y on the PRR) leading to the 1876 Centennial. The system used a Westinghouse steam-driven air pump and applied air at 20 psi to operate track switches and banner signals. The system was somewhat inefficient since the air admitted at the machine was required to charge the entire line to the operating device as well as the operating cylinder for that device. The Union Switch & Signal Company acquired the patents for this system in 1891.

Caleb Harlan Jackson, most often written as C. H. Jackson, manufactured the interlocking apparatus for the Toucey & Buchanan Company in his factory at Harrisburg, Pennsylvania. He later established the Jackson Manufacturing Company to additionally manufacture switches, frogs, crossings and switch stands, and even wheelbarrows. He was President of his company, Treasurer and Manager of the Toucey & Buchanan Company, and in 1879 became General Agent of The Union Electric Signal Company. As principal representative of these companies, he was able to provide clientele with an arrangement in which each company could complement the others in the case where their products were involved in signaling devices installation. The interests of the Jackson, Toucey & Buchanan and Union Electric Companies were consolidated in 1881 in The Union Switch & Signal Company. Jackson became General Manager of the combined organization, and remained active in the management of the company for many years. He was one of the incorporators of the Electric Company, and represented George Westinghouse in many corporations in which Westinghouse had a financial interest.

The first rail made its way into South Africa in 1860 when the Natal Railroad Company ran the "Natal," a small standard gauge well tank on June 26, 1860.

South Africa's first rail was installed in 1860, and was acquired by the Natal Government in 1877. The first railway in the region which is now called Pakistan was a 105-mile line from Karachi to Kochi, opened in 1861. New Zealand opened its first steam railway from Christchurch to Ferrymead in 1863, followed by Ceylon and Japan. Japan's first railway from Yokohama to Shinagawa opened June 1872 and the track miles expanded quickly. From 1880 to 1890, mileage grew from 98 to 1,459. By 1970, Japan's total of 16,953 miles exceeded Great Britain's 13,261 miles.

China shied from railway development until 1876, when a 20-mile line was opened from Shanghai to Woosung. After a fatal accident occurred, the Government bought railway, tore it up and dumped the remains on the island of Formosa. Their first permanent railway, the Tongshan-Hsukuchuang line, opened in 1880, with the introduction of steam traction in 1883. Chinese railway mileage grew quickly from 1,458 miles in 1900 to 21,750 miles in 1970.

Burma's first railway was a line between Rangoon and Prome, opened in 1877. By 1875, American railroads comprised over 70,000 miles, making it the largest railroad network in the world.

In 1890, Massachusetts' Board of Railroad Commissioners complained that the "lack of uniformity on different roads in rules governing the train service is a definite danger." It also noted "the alarming diversity of (signal) practice," observing that "the arrangement of lanterns which means safety on one line means danger on another."

The A1 "Terrier" Class locomotive, "Freshwater," in the Southern Railway livery at the Isle of Wight Steam Railway. The Isle of Wight is England's largest island.

SMALL, BUT REELING WITH RAIL!

Oddly enough, the quaint Isle of Wight, off the coast of Hampshire, England, measures only 20 miles east to west and 13 miles north to south, yet it boasted 42.25 miles of railway operated by seven separate companies, beginning with its first 4.5 miles of rail in 1862. Today, all that remains is a short line from Ryde to Sandown.

George Westinghouse

The Man Behind the Switch

If George Westinghouse Junior grew up in a present day American school, he would probably have been designated as having attention deficit issues. He wasn't attentive in school, and in those days, one was more likely to be perceived as too lazy rather than too intelligent. Even when he attempted college, the school informed him that there was nothing they could do for him, so he left.

Westinghouse was obsessed with how mechanical things worked. When he was 20, he watched a wrecking crew pry derailed cars with levers and jack them up before lowering them into place. He announced that a separate pair of rails could be clamped to the track on an angle, and aligned to the wheels of the derailed car, and an engine could replace the car onto the track. A friend suggested that he design this device and sell it to the railroads. So he did just that.

"Speed is limited by control. Control means that the power to stop must always exceed the power to go."

His father didn't approve of him heading into unknown manufacturing territory, and discouraged him time and time again with the same argument – "Do what you know, we don't know the rail business." Such resistance from one's own family might have discouraged others, but Westinghouse Jr. was just stubborn enough to ignore George Westinghouse Sr's advice and prove him wrong.

Westinghouse was an impetuous man who looked at problems and resolved to create a solution. His inventive nature was the impetus to bring his drawings to life. In 1869, he received his first patent for a locomotive air brake that revolutionized railroading, allowing trains to travel safer at faster speeds. When he determined that rail safety in the United States was in a dismal state, he designed an electro-pneumatic system to help train crews avoid accidents, and soon after built his first interlocking system.

He never sat on his laurels. As soon as one invention was in production, Westinghouse had several more up his sleeve. He thrived on experimentation, which led to inventions of a friction draft gear to control railway cars, large gas engines, an automatic telephone system and air springs (the forerunners of shocks) for automobiles. Westinghouse was the first Pittsburgher to drill for natural gas, and he did so on his own estate at Homewood. He built machines to generate and distribute AC electric current, apparently enjoying the "battle" with Edison on the AC versus DC debate that raged between the two men for years. If someone built a device first, he bought the patents and frequently hired the men who invented the apparatus to work for him.

Westinghouse was laconic, which often gave him the label of being cold, domineering and harsh, but those who knew him said that beneath his stern exterior was the kindliest spirit and a sympathetic heart. He built modern homes for employee rentals, and established the Saturday half-holiday. He kept men employed when there was insufficient work, and offered them clubs and societies for self-improvement and social activities. His school allowed female employees to study stenography, typing, cooking, sewing and household art or music at his expense.

He opened his home for Thanksgiving dinner to employees until the event's popularity outgrew the space, at which point he gave each employee a turkey for the next 30 years. When Westinghouse learned some households were getting so many turkeys that that the families were raffling off the extra fowls to neighbors (one father had seven single sons working with him), he used the "turkey budget" to start an employee pension fund – becoming one of the first American industrialists to do so.

In 41 years he built up 30 corporations of which he was President, holding $250 million of capital and employing 50,000 men. He received honorary degrees from Pittsburgh's Union College and The Koenigliche Technische Hochschule of Berlin. France took him into her Legion of Honor, King Humbert decorated him with the Order of the Crown of Italy and King Leopold II of Belgium personally pinned upon him the highest regal honor. He belonged to many organizations, including the Royal Institution of Great Britain, ASME, AIEE, Metropolitan Museum of Art and the Japan Society.

Westinghouse died of heart disease in 1914 at 68 years of age. He left $50 million in assets to his family and close associates, but more importantly, he left a legacy for the world that continues to this day.

27

The Union Switch & Signal sales book for 1890 was a catalog that contained product descriptions accompanied by detailed drawings.

Photo credit: Joanne L. Harris

3 Birth of The Union Switch & Signal Company

Alfred Waud depicted this interior view of a first-rate Pullman Palace car in a 1869 issue of Harper's Bazaar.

In 1830, America had 40 miles of steam locomotive track. By 1880, the railroads stretched 90,000 miles across the entire continent. First class accommodations, such as Pullman Palace cars, specialized restaurant railcars and sleepers separated those with silk-lined pockets from those who had skimped to be able to travel to their destination; those same who would try so desperately not to fall asleep on the shoulder of a stranger. There may have been a rebirth in the economy in the 1880s, but its short revival confirmed that the railroads were largely overbuilt, financially famished and fighting for business between scores of struggling commercial ventures masked as railroad corporations. Some would survive the rate wars; many would perish or fall into the hands of wiser businessmen.

Rate wars spurred high increase in demand for train travel and cargo shipping, bringing along with it high increase in injury and death, because American railroads believed at the time that it was more financially feasible to pay the lawsuits and other damages than to pay the cost of installing and maintaining crossings. This conception would soon be disproved. In the five years between September 30, 1877, and September 30, 1882, New York railroads killed or injured 264 individuals, which generated lawsuits and other damages averaging $358 each (about one year's annual earnings in 1880).

By the 1880s, some American main lines were carrying more than one hundred trains every day. Such density put enormous pressure on single-track railroads. The increased number of potential meeting points magnified the possibilities for human error that could cause train collisions. Automatic blocking on single-track lines, though, was technically complex. Trains must be stopped in both directions. Additionally, the early signals did not function well. The electromagnets struggled to transmit power long distance to the signals, and the signals were difficult to see. By the 1880s, battery-controlled pneumatic signals that used compressed air to power a semaphore were installed, but the cost was

"A remarkable railway wreck" occurred on January 17, 1888, on the Rochester line of the New York, Lake Erie and Western Railroad. Train 18 managed to get ahead of Train 107, and the dispatcher confused his orders. Both trains were running at high speed and met on a sharp corner.

Throughout the 130-year history of The Union Switch and Signal Company, the company underwent numerous official and unofficial name changes, including a period when it was officially known as Westinghouse Air Brake Company. For the purpose of this book, the official name changes will be referenced, but the name throughout the book — until it's complete name change to Ansaldo STS USA — will be designated as Union Switch & Signal.

rather prohibitive. Later signal designs were powered by more affordable, electrically-activated CO_2 cartridges. Finally, about 1900, the invention of small electric motors to run the semaphores made electric signal systems practical.

It was during this time that George Westinghouse set a new strategy in motion – to enter the interlocking switch and signal business as a major player. His overwhelming success with the airbrake strengthened both his reputation and his financial standing. He had been watching these safety equipment businesses both in the United States and England.

The following events describe a strategic path, which would serve Westinghouse well in the development of his business. It was not by chance that he was present at the meeting in which Toucey and Buchanan signed the agreements with Saxby and Farmer for their patents in 1887. Nor was it a fluke that he bought stock in the company and became its President. This was a well-planned, well-executed series of business deals on the part of Westinghouse, whose business acumen would quickly place his company in the forefront of the rail industry.

Westinghouse recognized that rail traffic signaling in the United States remained in an archaic state. Several leading American railroads had invested in the English block signaling system for their busier rail yards, junctions and crossings; but the vast majority of the nation's tracks remained unprotected. Signaling reduced accidents and kept traffic moving with reliable frequency. The railroads divided their tracks into sections of blocks from a half-mile to four miles in length. A signalman raised a "danger" signal when a train approached and kept it up until the signalman at the station ahead telegraphed an "all clear" for the block.

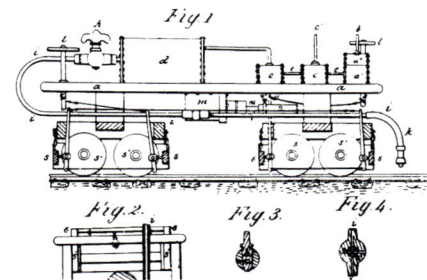

George Westinghouse's 1869 patent for an Improvement to the Steam Power Brake.

"If some day, they say of me that with the air brake I contributed something to civilization, something to the safety of human life, it will be sufficient."

– George Westinghouse

Westinghouse was unsuccessful in using electrical circuits to control pneumatic valves on trains, but electrical power could conceivably work if an air pump and reservoir were installed close to the switch or signal. The signalman could then throw a switch to send electrical current through a wire to activate an air valve near the track. This obviated the need to install or maintain lengthy air lines. The design of his electro-pneumatic system used the metal rails to carry the electrical current, which reduced the amount of wiring required and served as an automatic safety feature. An open switch or broken track interrupted the electrical circuit, which in turn raised an automatic danger signal. If either the pneumatic equipment or the electrical circuit broke, the trouble could be detected and repaired in a fraction of the time required to overhaul an all-pneumatic system.

Union Switch & Signal workmen in a Swissvale storeroom in 1888.

On the corporate front, George Westinghouse and Henry Snyder were elected directors to fill vacancies in William Robinson's The Union Electric Signal Company in February 1881, and Westinghouse was also elected President. As part of the bargain, he was able to purchase 10,000 of the company's total 18,016 shares of stock, which constituted a controlling interest in the company. At the March 17th Directors' meeting, it was voted that the manufacturing and office facilities were to be moved from Boston to Pittsburgh on or about May 1, 1881, or as soon thereafter as possible. A week later Ralph Bageley was elected Vice President and Henry Snyder was elected General Agent. A committee of three (George Westinghouse, Ralph Bageley and J. Gardner Sanderson, a past vice president of the Union Electric Signal Company) was appointed to devise plans to consolidate The Union Electric Company and the Interlocking Switch and Signal Company, by increasing the capital stock if necessary.

On March 21, 1881, all of the Toucey & Buchanan company patents were assigned to George Westinghouse, after which he signed them over to the Interlocking Switch and Signal Company on March 28. These patents were re-assigned to The Union Switch & Signal Company on or about May 1, 1881.

E. L. Frothingham, Jr., Treasurer of The Union Electric Signal Company, wrote a letter to "Dear Sir" (presumed to be Franklin L. Pope, then on the Board of Directors) on April 14, 1881, that a special meeting of the stockholders of the company was held at their office in Hartford Connecticut the day prior. The Board, he related, unanimously voted to change the Corporation name to The Union Switch & Signal Company. They voted to increase the capital stock from one million to $1,500,000 with the increased stock ($500,000) "to be used in whole or in part in payment for the assets and property of the Interlocking Switch and Signal Company." George Westinghouse was authorized to hold the stocks as trustee. They also voted to authorize

GEORGE WESTINGHOUSE, JR. President.
C. H. JACKSON, General Manager.

HENRY SNYDER, General Agent.

RALPH BAGALEY, Vice Prest and Treas.
ASAPH T. ROWAND, Secretary.

THE UNION SWITCH & SIGNAL CO.

Pittsburgh, Pa. *Aug 3d 1882.*

For value received, I hereby guarantee the due payment of any and all promissory notes or their renewals, drawn by The Union Switch & Signal Company, and endorsed by C. H. Jackson and Ralph Bagaley, or their successors, to an amount not exceeding Forty thousand dollars. ($40,000.00)

Geo. Westinghouse Jr

To
J. M. Gordon, Esq.
Cash'r People's Nat Bank
Pittsburgh Pa.

the directors to receive the assets and property of the Interlocking Switch and Signal Company of Harrisburg, Pennsylvania. The treasurer was authorized and instructed to purchase all 1,000 shares of the capital stock of the Interlocking Switch and Signal Company now owned by Saxby and Farmer of London, England, for 6,500 British pounds sterling. The company acquired all the property including the valuable patents of the Interlocking Switch and Signal Company when the two organizations were consolidated on May 1, 1881.

The Interlocking Switch and Signal Company had a total of 5,918 shares; 1,000 were owned by Saxby and Farmer of London, 4,102 were owned by the company and 816 were contributed by stockholders of the Interlocking Switch and Signal Company. The Board entrusted 4,168 shares to Westinghouse as trustee to sell 1,750 shares at "the highest price possible." The stocks were sold to reimburse the company's funds for the stocks purchased from Saxby and Farmer, and to buy a manufacturing property in Pittsburgh known as the Bidwell Plow Works, with buildings and iron-working machinery, which Ralph Bageley had purchased from a Sherriff's sale.

C. H. Jackson became the General Manager in 1881. He was the man who put the Union Switch & Signal deal together for Westinghouse, and Jackson held the position from 1881 to 1886. He assembled an executive team that included Frederick Gürber, Chief Engineer; Oscar Gassett, Electrical Engineer; Charles R. Johnson, Signal Engineer and New York office manager; and Asaph T. Rowand, Secretary. Both Gürber and Rowand had worked for Jackson at The Interlocking Switch & Signal Company.

Early in 1881 Westinghouse obtained patents for a switch and signal apparatus and for a block signaling apparatus, covering hydro-pneumatic and electro-pneumatic devices. Initial sales were sluggish, but he slowly convinced the railroads of the advantages of his new traffic control devices. A few railroads purchased equipment and installation based on his excellent reputation with the air brake. The growth of the business was slow, but as the railroad industry developed, signaling devices gained the same full acceptance that the air brake had achieved. On November 11, 1882, the company was chartered in Pennsylvania as "The Union Switch and Signal Company." By 1884, there were about 100 Union Switch & Signal interlocking systems with a combined total of 2,100 levers installed throughout the United States. This compared to only eight interlocking systems installed by other manufacturers, six of which were built by the Pennsylvania Steel Company. The railroads were more attracted to the use of switch and signal interlocking where railroads crossed each other, in good measure due to state actions.

Westinghouse investigated the British innovation of interlocking, in which a system used huge levers to control the railroad switches and

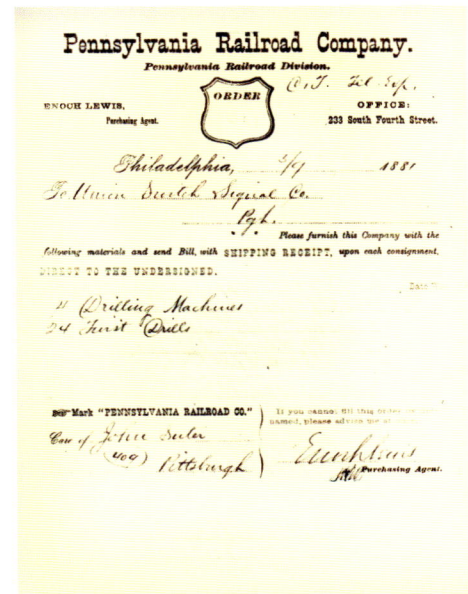

Union Switch & Signal received its first order from the Pennsylvania Railroad Company (PRR) on May 9, 1881. It signified the beginning of a tightly knit business relationship that would last throughout the life of the PRR.

Westinghouse's automatic block system introduced on the Central Railroad of New Jersey.

Opposite: An 1882 letter of guarantee from George Westinghouse to T. M. Gordon, Esquire, at the People's National Bank in Pittsburgh on Union Switch & Signal's original letterhead.

The Union Switch & Signal electro-pneumatic interlocking machine installed at Boston Southern Station was recognized as a profitable investment at busy crossings after weighing the cost of stopping trains versus installing interlockings.

Shelby M. Cullom

signals in a block that were assembled into a single machine and interlocked with each other, so conflicting signals could not be given. This meant if someone were to pull levers at random, they could stop traffic but they could not cause a collision.

Westinghouse had been experimenting with a device called "the button system" for electrifying the horse car. Iron pins were buried in pairs at intervals between the tracks to supply electric current for the trolleys. Each pair of pins connected with electrical conductors which led to electromagnetic switches alongside of the track. The switches were linked with a power line laid in a conduit beneath the pavement. The pins were inert and harmless except when two iron bars, projecting like tuning fork prongs from the bottom of a passing trolley, made contact with them.

Throughout the 19th century most Americans wanted railroads. State and local government wanted safety at grade crossings, but they usually required only whistles and bells before a crossing rather than gates or watchmen. In 1882, New York State had 6,901 crossings; only 703 of which had gates or flagmen. After the rear-end collision of the New York Central at Spuyten Duyvil, the Railway Gazette reviewed Britain's progress and once again prodded the United States to block their lines as well. In 1884, the American Railway Association realized the need to effect change upon railroad safety, and adopted the Standard Code of train rules. Not all of these rules, though, would be successful.

By 1884, there were about 100 Union Switch & Signal interlocking systems with a combined total of 2,100 levers installed throughout the United States. This compared to only eight interlocking systems installed by other manufacturers, six of which were built by the Pennsylvania Steel Company.

Rule 99 of the Standard Code required the conductor to send a flagman to protect his train whenever it stopped on the road. Thus the safety of two trains was left in the hands and judgment of the least experienced, worst paid crewmember. Railroad Gazette Editor Arthur Wellington penned his feelings about relying on Rule 99: "Our railroads have been trying this experiment now for nearly sixty years, only to prove by each year's melancholy results that when flagging is depended upon there is no other word than 'fail.'"

In the early- to mid-1880s, numerous states transformed interlockings from a money loser into a profitable investment at busy crossings when they weighed the cost of stopping a train at every crossing versus buying, installing and maintaining an interlocking. Around 1900, one Canadian and Northwest Railway signal engineer, J. A. Peabody, estimated that it cost an average of 45 cents to stop a passenger train. Heavier freight trains consumed even more coal, which was the highest expense in stopping a steam locomotive. He set that figure into a table with the examples of 14, 20 and 25 trains daily. When a crossing had to handle even 14 trains per day, the total cost of the interlocking not only saved time, but money. Fourteen trains at

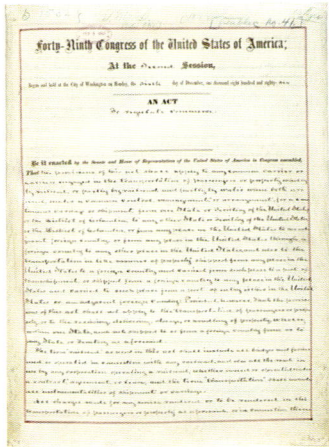

U. S. President Grover Cleveland signed the Interstate Commerce Act on February 4, 1887, which required interstate rates to be "reasonable and just."

45 cents times 365 days a year equals – $2,299.50. At that time, the railroad could buy "an interlocking plant for a single track crossing, where 16 levers would be required" and hire a day and night tower man, at the estimated annual cost of $1,968 per year, including the cost of the interlocking, interest, depreciation, maintenance and operation.

In 1885, a special Senate committee headed by Shelby M. Cullom, ex-governor of Illinois, conducted an investigation of railroad practices. The final report of the Cullom Committee in January, 1886, listed all the familiar abuses of unreasonably high local rates; discrimination between persons, places and types of freight; special secret rebates and drawbacks; passes; watered stock that caused excessive capitalization; and managements that were extravagant and wasteful. The report recommended the creation of an independent commission to regulate the nation's railroads.

A second event in October 1886 also indicated that federal action was near. In the Wabash, St. Louis and Pacific Railway v. Illinois case, the Supreme Court ruled that a state could not regulate shipping rates passing beyond its own borders. The federal government would determine the regulations of interstate commerce. Cullom's Senate bill and John H. Reagan's House bill were reconciled into the Interstate Commerce Act and the formation of the Interstate Commerce Commission (ICC), which President Cleveland signed on February 4, 1887. The ICC, albeit rather vaguely, required all interstate rates be "reasonable and just" and prohibited rebates, drawbacks and pools. (These pools created opportunities for rate clerks and traffic men to collaborate on rates without the benefit or application from Congress law.) It further required the railroads to publish their rate schedules "in every depot or station,"

Time Zones

By 1872, there were already ten official prime meridians in use, complicating the work of the shipping industry. The United States Navy and Britain (plus her colonies) used the Greenwich charts, but ten percent of the world did not. A number of men devised proposals to reduce the number of local times to benefit scheduling, including Professor Charles Dowd, Principal of Temple Grove Ladies' Seminary in Saratoga Springs, New York (now Skidmore College). Dowd proposed a Five-time-zone system for North American railroads. Each zone covered 15 degrees of longitudes, with Greenwich, England serving as 0 degrees. William F. Allen, secretary of the American Railway Association, heard him out but dismissed the plan.

On April 8, 1883, 50 grand-trunk railroads managers voted to accept William Allen's proposal for standardizing time into four zones – Eastern, Central, Mountain and Pacific. A fifth zone, to include Canada's maritime provinces would soon be designated "Intercolonial." On November 18, 1883, Railroad Standard Time was established across North America. That day became known as "the Sunday of Two Noons," as towns on the eastern edges of each zone had to turn their clocks back half an hour, creating a second noon. Within a few days, 70 percent of North America's schools, courts and local governments adopted "railroad time" as their official standard.

In October 1884, President Chester A. Arthur held the Prime Meridian Conference in Washington, D.C. to set standard time for the world. American and British delegates pushed 23 other nations into accepting Sir Sanford Fleming's time zone plan, with Greenwich as the prime meridian. Allen and Fleming were touted as the creators of modern standard time, while Dowd, to whom much of the credit was due, was not even invited to the conference, nor was his name mentioned. Dowd died in a railroad crossing accident in Saratoga Springs in 1904. Congress wrote Standard Time into law in 1918.

Westinghouse Electric Company's 1888 Catalog – George Westinghouse developed the incandescent technology and was building lamps at Union Switch & Signal for nine years before the Westinghouse Electric Company illuminated the Chicago World's Fair with its two-piece stopper lamp.

which was rarely observed, and to file them with the federal government. Although the Act of 1887 made pools illegal, only the rate clerks had the financial comprehension required to comply with the published tariff rates. Traffic men would struggle to keep their pools together.

At the same time that Union Switch & Signal was building its railroad equipment business, George Westinghouse was already keen on entering another new field – lighting. While it is widely known that Westinghouse Electric Company started in July 1886, it is a scarcely known fact that in 1884, George Westinghouse hired American physicist William Stanley as his Chief Engineer to run experimental work on the incandescent light at the Union Switch & Signal plant. Thus the premise that the Westinghouse two-piece stopper lamp that illuminated the Chicago World's Fair (Columbian Exposition) in 1893 was his first attempt in the field of lighting is incorrect. Westinghouse had been manufacturing incandescent lamps at Union Switch & Signal for nine years before the Fair.

Stanley offered great ingenuity and promise to the Electric Light Department at Union Switch & Signal. He had previously devised an improved method of exhausting incandescent lamp bulbs in 1881, and with a partner, E. P. Thompson, Stanley developed a method of treating silk thread to manufacture incandescent

William Stanley

lamp filaments, which the Sawyer-Mann Company then used in their incandescent lamps. (Westinghouse acquired the Sawyer-Mann Company in 1884.) During 1884 and part of 1885, Stanley installed and ran a fully-equipped factory to produce incandescent lamps and the primitive track circuit equipment that Robinson had designed at "the Lawrenceville Works of the Union Switch & Signal Company." While it is not certain, it is probable that "Lawrenceville" refers to the neighborhood just north of the Garrison Alley address, and thus the reference is most likely the Garrison Alley works.

Stanley produced the first alternating current transformer during his tenure at Union Switch & Signal. Westinghouse imported equipment that used the Gaulard & Gibbs system in the spring of 1885, and having determined its advantages and disadvantages, Stanley built a better system. Due to health issues, Stanley moved in October of that year to Great Barrington, Massachusetts, where he built and equipped a complete laboratory for experimental work. He then designed the form of induction coil, or, as he called it, a "converter," and a self-regulating system of alternating current distribution – or transformer – that the Westinghouse Electric Company would come to use.

Stanley demonstrated his theory by equipping a lighting plant in Great Barrington in the spring of 1886, which may have been the first successful alternating current, electric lighting plant in the United States. By this time, Westinghouse had filed an application to charter the Westinghouse Electric Company, and the design and manufacturing activities were all transferred from Union Switch & Signal to the new firm. In 1893, Union Switch & Signal manufactured and widely advertised their "Stanley and Thompson lamp" after buying the patent of the joint invention by William Stanley Jr. and Edward P. Thompson.

George Westinghouse arranged for the Westinghouse Electric Company to take over the Garrison Alley shop in 1886 and move Union Switch & Signal's operations to Swissvale, about eight miles east of Pittsburgh. That same year, Charles R. Johnson became General Manager of the company.

Johnson, a native of Higham Ferrers in Northamptonshire, England, was born in 1851. He was educated at Dr. Pinches' Academy in Kennington, London, and went to work afterwards in the drawing office of London's City Architect department. When he was 23, he

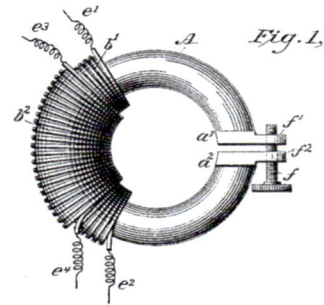

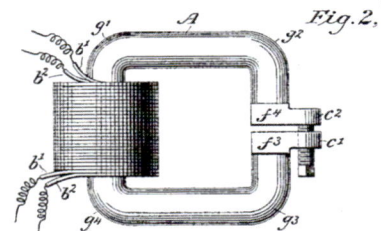

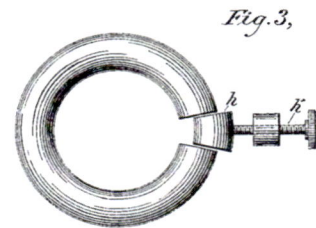

Figures 1, 2, and 3 detail the 1886 induction coil patent for William Stanley's transformer (final drawing), which was first used in the lighting plant at Great Barrington, Massachusetts.

The Union Switch & Signal plant in Swissvale, Pennsylvania, as it appeared in 1892.

found employment at Stevens & Sons, a railway signal manufacturing firm in London, where his father already worked. His uncle, Henry Johnson worked at Saxby & Farmer, and in 1875, he found a job there. Within a few years, he was overseeing the installation of signals in England, Ireland and France. According to his obituary in the October 1893 issue of *American Engineer*, he went to India as Saxby & Farmer's representative in 1880, but contracted jungle fever, from which he never fully recovered.

In 1881, the Pennsylvania Railroad Company was facing issues with the crossing of the PRR line with the Central Railroad of New Jersey at Elizabeth. They requested Saxby & Farmer to send them an advisor. The British firm sent Charles Johnson, their most competent signal engineer. Shortly thereafter, he began working for Union Switch & Signal as a contracting agent for them in New York. Johnson became General Manager for Union Switch & Signal in 1886, but his tenure there was brief. George Westinghouse dismissed him after a nasty dispute during which Charles Johnson demanded that Westinghouse give him an increase in salary, a cash payment, a large block of stock in the company, the position of Vice President, and virtual control over all company policies. Westinghouse shared this information in a statement with the "Pittsburgh Dispatch" on December 9, 1888.

Charles Johnson

These demands, accompanied by the threat that the company would fail if the Johnsons (he and his uncle) left, drove Westinghouse to promote Henry Snyder to Managing Director, with authority over the General Manager. Westinghouse and Johnson came to an agreement that the 10-year contract would be severed the following July, but an early dismissal of Charles and the resignation of Henry sent them both packing. Charles Johnson and his uncle Henry relocated to New Jersey, where they organized the Johnson Railroad Signal Company in Rahway, of which Charles was President and General Manager. There they engaged in the interlocking switch and signal business, amidst a litigious relationship with Union Switch & Signal.

Snyder discovered that copies of all Union Switch & Signal drawings had been made, beginning in November 1887. The company took action against Johnson's company and the sheriff discovered a bundle of prints from Union Switch & Signal. Charles and Henry Johnson denied any connection with the theft, although Charles stated that a former, disgruntled employee of Westinghouse had taken them. Suits and countersuits between the two companies were filed for an array of claims: patent infringement, breach of contract, breach of trust, and so on.

Henry Snyder was the Managing Director from 1888 to 1890. Little was documented on Snyder, who became an affiliate of George Westinghouse in 1881. Early on, he became the General Agent for Union Switch & Signal, and held that post when Charles Johnson was working as the New York representative for the company. After the skirmish with Johnson, Snyder was promoted and held the position until he died in 1893. Edward H. Goodman then filled the position of General Manager.

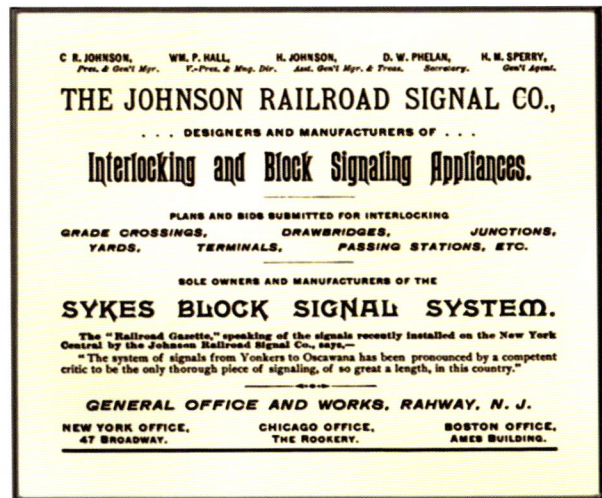

Charles Johnson and his uncle Henry Johnson relocated to Rahway, New Jersey, where they organized The Johnson Railroad Signal Works.

In 1888, Electrical Engineer magazine, Volume 7, reported that Union Switch & Signal had "made a departure in the manner of supplying interlocking and signaling apparatus." The company had previously only contracted to manufacture and sell products, but now they would bid "for material with superintendence," meaning, they would install everything they sold. If a railroad only wanted the finished product, Union Switch & Signal would then offer two bids – one for manufacturing only and a second bid for installation.

The decade was full of innovation, litigation and reorganization, and Union Switch & Signal was underway. America was catching on to the benefits of railway signaling, but England still maintained a strong lead. The British Parliament passed The Regulation of Railways Act of 1889, which authorized the Board of Trade to require railway companies with passenger railways to adopt the block system of signaling, provide interlocking of points and signals, and to use continuous automatic brakes on the trains. It also addressed the need for a reporting system of all safety-related personnel who worked more than a specified number of hours. All passengers now had to show their tickets to prove they were indeed paying passengers, and they would have to pay a penalty for not having one. In the United States, similar changes would soon follow.

The prominent inventor,
William Robinson, in 1888.

4 The Depression, the Coup and the Competition

The financial distress of American railroads throughout the 1880s could not hold up through the depression of 1893 to 1896. Financial reorganization and consolidation of railroad companies was impending, as an increasing number of bond defaults forced bankers to remove railroad management and place themselves as trustees for the bonds. As the bonds were usually secured by first mortgages on the railroads' physical assets, bank leaders found themselves more and more involved in railroad affairs. It was an inescapable reality that after the Panic of 1893 they would become more involved in the day-to-day management of many railroad companies.

The depression hit at the same time America was immersed in transformation – shifting from an agrarian society to an industrial society, perhaps similar in nature to the advent of the computer generation. Advancements in machines, electricity and transportation all played a part in that transformation, as people were awed by new inventions that would forever change their lives. In the midst of this transformation, according to Westinghouse biographer Henry G. Prout, Westinghouse spent half of the 1880s adding a patent per month to his resume, and that impressive pace of innovation lasted into the mid 1890s. In that 11-year period, he applied for 134 patents in half a dozen different arts – perhaps the most creative years of his life.

In 1892 only an estimated 3,000 to 6,000 miles of railroad, or less than 4 percent of the total, were blocked.

Westinghouse delved into many areas of rail safety and electrical power in this era. He was adverse to using electrical devices in close proximity to, much less upon, railway tracks. The railway men of the day shared this aversion. Consequently, his early interlocking efforts incorporated hydraulic pressure to apply power and pneumatic pressure to set the interlocking in operation to handle the switches. He actually avoided using electricity for switch operation until 1890. By the following year, Union Switch & Signal power interlocking design changed from a hydra-pneumatic system to the electro-pneumatic system. The new system used magnet valves to control compressed air for the operation of both interlocking signals (the

The Panic of 1893 hit the New York Stock Exchange, and drove bankers to take over as bond trustees for the railroad companies as one after another defaulted on their mortgages.

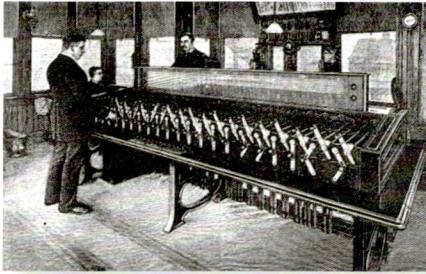

Union Switch & Signal's Electro-pneumatic interlocking system at Pittsburgh, Pennsylvania, replaced many of Westinghouse's installed mechanical interlocking machines over time.

Major Edward Harris Goodman (above) worked his way up the ladder at the Pullman Palace Car Company, shown here in 1900.

semaphores) as well as the switch points. The Company, with about 200 shop workers and 200 or so more in the field, built and installed 21 of Westinghouse's first hydra-pneumatic interlocking machines between 1883 and 1891, all of which were replaced in time by electro-pneumatic systems.

While Westinghouse was president of all his companies, he made every effort to hire the best men to run the daily operations of those companies. One such choice was Major Edward Harris Goodman, who for 18 years had worked his way up the ladder at the Pullman Palace Car Company until he became a second Vice-President and General Manager in 1880. He left the Pullman company to work as the Assistant General Manager and General Agent (a sales manager) for Union Switch & Signal. When Henry Snyder (the General Manager who had been with the company since its inception) died in 1890, Goodman was promoted to his position. Later that year, the office of Secretary-Treasurer was moved from Pittsburgh to Swissvale. Caleb H. Jackson resigned as Vice President (but remained on the Board), and Goodman was elected to succeed him.

It was during this time that the "executives became restless." Union Switch & Signal as a public company had, for a number of years, solicited proxies from the stockholders that authorized either George Westinghouse or Asaph T. Rowand, the Secretary, to vote their stock at the shareholders' annual meetings. George Westinghouse was in financial difficulty due to the scarcity of ready money in the country, and he traveled to New York City to discuss refinancing for his Electric Company. Thus, he was unable to attend the annual shareholder's meeting in Pittsburgh on March 10, 1891. Rowand exercised the right to vote the proxies. Seven directors were elected for the succeeding year. Rowand voted in representatives of "Eastern" interest, principally a group of Boston-based shareholders who had been associated with the original Union Electric Signal Company. Six Eastern directors were elected, along with Rowand. The former board included Westinghouse, Goodman, Jackson and Robert Pitcairn, who was also General Agent and Superintendent of the Pennsylvania Railroad. Upon learning of the "coup," Pitcairn protested on behalf of all the ousted board members, objecting to the legality of the election.

> **Rowand and his Boston cronies may have felt that Westinghouse and his genius as an inventor would never give the business the serious nurturing that it required to make the business flourish as they thought it could.**

At the Directors organization meeting held the same day, Rowand was elected President, Sigourney Butler of Boston became Vice President, and E. H. Goodman was elected General Manager. Rowand was authorized "to institute such means of reform as will result in a more economical administration of affairs," and was authorized "under advice of counsel to initiate such suits as is necessary to assure that all assets of the company are secured, and wrongs to the company righted," according to minutes of the meeting. Thus began the three-day reign of Rowand.

Rowand's motivations will never been known, but one author offered a logical possibility that Rowand and his Boston cronies may have felt that Westinghouse, despite his genius as an inventor, would never give the business the serious nurturing it required to make the business flourish as they thought it could. Regardless of his intentions, as soon as Westinghouse learned the results of the meeting, he took the first train from New York to Pittsburgh and conferred with his associates, arranging for another directors' meeting on March 14. This meeting was held in Philadelphia with four of the new directors present; but noticeably, neither Rowand nor Butler was invited.

One of the directors present offered the resignations of Rowand and Butler from their respective new offices, which the Board immediately accepted. E. H. Goodman was elected President, and Rowand was elected to the position of Vice President with limited powers and an insignificant salary. It was reported that when Rowand was informed of the Board's action, "he fell unconscious to the floor of the Duquesne Club and was later removed to his home where he lay in a critical condition, said to result from a stroke of apoplexy." In May 1891, Rowand submitted a letter to the board resigning his position as vice president. The resignation was immediately accepted.

At the company's annual meeting on March 8, 1892, Westinghouse represented the majority of the stockholders' interest. Seven new directors were elected, including Westinghouse, E. H. Goodman, E. Scott Fitz and four prominent Pittsburgh men: A. M. Byers, Thomas Rodd, James H. Willock and William McConway. Westinghouse was elected President, E. H. Goodman was elected Vice President with the active company management delegated to him, and James Johnson was named Secretary-Treasurer. Westinghouse remained President of the company until his death in 1914. Henry G. Prout, former editor of the Railway Gazette, replaced Goodman after he retired in 1893.

Colonel Henry Goslee Prout, former editor of the Railway Gazette

Union Switch & Signal may have had competition in-house for leadership, but during the first few years of its existence, it had relatively little, if any, external competition in the fields of automatic block signaling, mechanical interlocking and power interlocking. This situation began to change about 1889 to 1890 when the basic patent protection of Robinson's close track circuit expired. The Hall Railway Signal Company reorganized as the Hall Signal Company, and became active in the block signaling field.

America's Greatest Railroad — for Rear Collisions

By the 1890s, the growing density of rail traffic was rapidly outstripping the ability of existing tracks to keep trains safely separated. The New York Central line near Hastings, New York, ran 168 trains a day on its main line (without blocks), three times the density of the little 44-mile Providence & Worcester, which had used blocking since 1881. The P&W had been collision free since then, while the Central had collisions "by the dozens." Angus Sinclair, editor of Locomotive Engineering, observed that the Central's publicity termed it "America's Greatest Railroad," and suggested they add " – for Rear Collisions," which he claimed "are of almost daily occurrence." He chastised them for allowing trains to follow each other in line-of-sight, with "sixty-car freight cars without air brakes" mixed with locals and "sixty-mile-an-hour expresses."

The Johnson Company (and later the National and Standard companies) entered the field of mechanical interlocking. John D. Taylor and F. M. Dodgson became active as developers of the electric and low-pressure pneumatic systems, respectively.

Like any other emerging field in history, it would not be a peaceful competition as industry leaders fought for supremacy and market share. Although the basic track circuit patents expired in 1889, Union Switch & Signal owned patents on essential improvements for commercially viable track circuit systems, including improvements for bond wires, distant signal control and an overlap system for signaling. The latter was an arrangement such that the control limits for a block signal overlapped and extended beyond the advance signal – that is to say, a signal to stop prior to the next block did not clear when the first train left the block, but rather, it continued to hold the following train until the lead train passed through a portion of the next block. This arrangement was deemed necessary particularly on downhill grades to provide sufficient braking distance via two stop signals behind each train.

In 1889, the Hall Signal Company in Garwood, New Jersey, offered the Hall wire circuit system using track treadles to control their banjo-style signals. They then developed the "normal danger" system of signaling. Union Switch & Signal sued on its overlap, distant signal control and bond wire patents, but all three were ruled to be invalid or not infringed upon by the Hall system. This established the Hall "normal danger" system as being outside of the patents of the Union Switch & Signal's "normal clear" system. A spirited controversy ensued between the two signaling companies. The Hall Signal Company had acquired a Buchanan patent on a relay arranged to overcome the dangers of lightning using the contacts of the Robinson track circuit. The relay provided for independent front and back contacts such that one set of contacts could still close even if the other set fused together. Union Switch & Signal sued them over this patent in 1894, and a decision was handed down in 1901, holding the patent valid and infringed.

In 1894, Union Switch & Signal introduced a banjo-type, enclosed disc signal to compete with the Hall Signal Company. Around 1897, they developed and introduced their Selectric semaphore signal, soon making the banjo and clockwork disc signals obsolete. Early forms of electric semaphores operated by means of mechanisms placed on the outside of the post, whereas Union Switch & Signal's signal design incorporated a mechanism at the bottom of the post that was connected by an electrical charge running through the post to the semaphore blade.

> In the 1890s the three-position semaphore went out of favor because it was not easy to arrange the mechanism so that the arm was accurately set. Sometimes the arm took up a position between stop and caution. It was replaced by the two-position signal with the arm falling vertically into the slot in the post, as before, to indicate 'clear.' The semaphore was not failsafe. It could be frozen in the "clear" position by snow or ice when in fact it should be in the "stop" position.

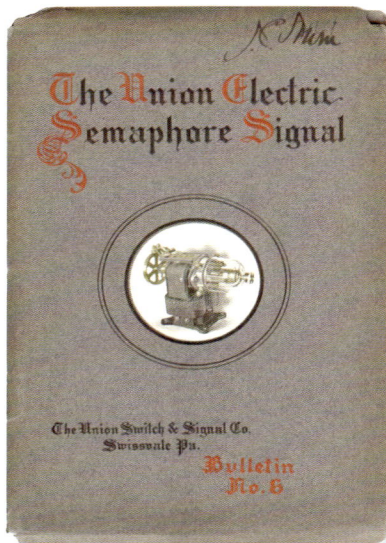

The Union Electric Semaphore Signal by The Union Switch & Signal Co., Swissvale, Pennsylvania Bulletin No. 6, November 1901.

During this same period, the Hall Signal Company developed an electro-gas semaphore in which bottled carbon dioxide operated the blades under the control of magnets governed by the track circuits. Union Switch & Signal started to follow their lead, but they quickly withdrew from this venture after the Hall Company filed suit against them. Not that it mattered in the long run. The electro gas signal never became popular and could not compete with the more successful electric semaphore. (In 1926, Union Switch & Signal took over the Hall Company and stopped manufacturing operations at Garwood.)

The Johnson Company (founded in 1888 by Charles Johnson) initially offered the only serious competition to Union Switch & Signal for mechanical interlockings. William P. Hall, who was at the helm of the Hall Signal Company, bought a controlling interest in the Johnson Company in 1892, but he never integrated operations of the two companies. Charles Johnson died in 1894, and a simultaneous financial panic brought the operations of the Johnson Company to a halt.

The National Switch and Signal Company, another entrant in the growing railway signaling business, acquired the assets of the then defunct Johnson Company in 1895, and along with the Standard Railway Signal Company that was created in 1896, began competing with Union Switch & Signal in the field of mechanical interlocking. In 1898, Union Switch & Signal, which then employed about 300 men in its shops, purchased the National Switch & Signal Company (with its 200 or so employees). They moved the machinery from the Easton, Pennsylvania,

The Hall Signal Company, with its factory at Garwood, New Jersey, was a hot competitor for Union Switch & Signal for many years.

The Union Switch & Signal plant in Swissvale as it appeared in 1890.

Many of the greatest railroad yards in the United States, such as the Delaware, Lackawanna and Western Railroad, began implementing the electro-pneumatic interlocking machine, and orders soon came in from abroad.

works into the established shops at Swissvale after building a large two-story brick and stone factory to accommodate the equipment.

New developments in interlocking continued to emerge, and just as quickly, several more companies were formed and merged. Between 1888 to 1889, John D. Taylor invented and built an electric interlocking machine featuring electrically controlled track switches. A little later the system was arranged to employ a "dynamic return" indication. The Taylor Switch & Signal Company, established about 1895, did not sell the machine, but when Taylor set up the Taylor Signal Company in 1900, the firm took over the assets of the original Taylor Company and began to push the sales of interlocking equipment. (Union Switch & Signal would later contract John Taylor to work for them.) Another inventor, F. M. Dodgson, also devised a power interlocking in the 1890s, which was developed by the Pneumatic Railway Signal Company. That firm later merged with the Standard Company and became the Pneumatic Signal Company. The Pneumatic Signal Company and Taylor Company consolidated in 1904 to form the General Railway Signal Company.

George Westinghouse and Jenns G. Schreuder (the co-inventor) received a patent for Union Switch & Signal's most elaborate and fully vital electro-pneumatic interlocking in February 1891. Schreuder later became Chief Engineer and one of the Vice Presidents of the Company. Many of the important railroad yards in the United States began using this design of interlocking, and it went into significant use abroad.

The companies all benefited from good timing. The railroads were burgeoning in the mid- to late-1890s and required exponentially larger electro-pneumatic interlocking systems to handle that growth. This was great news for Union Switch & Signal, whose products had produced sluggish sales – as with most entrepreneurial start-ups – up until this time. Their hard work was finally going to pay off. Over a course of about ten years, the company built and installed three of the largest interlocking systems in the world. The payoff began in the city of Boston.

Boston had a dilemma. Five railway companies were attempting to serve the city with five different depots, forcing passengers to haul their baggage between terminals. The state legislature granted a charter to the New Haven Railroad and the Boston & Albany to form a new corporation, the Boston Terminal Company. Their mission was to construct and maintain a union passenger station that would serve the five rail companies in South Boston.

John Pressley Coleman (aka J. P. Coleman), a self-educated engineer for Union Switch & Signal, was tasked to supervise the installation of a large interlocking machine at Massachusetts's Boston Terminal South Station on the Boston & Maine Railroad to operate switches and signals at Causeway Street. Coleman took with him B. H. Mann of St. Louis, Missouri, an MIT electrical engineering graduate whose exemplary year of work in the Union Switch & Signal Swissvale shops earned him an assignment in 1891 to work on the installation of automatic block signals on the Cincinnati, New Orleans & Texas Pacific. Mann became Superintendent of Signals for Union Switch & Signal on that railroad for seven years. When the project ended, he worked on the interlocking installation on the Boston Terminal (South Station) under Coleman. After just two years of construction, South Station, the largest railroad station in the world, was completed and the first train left the station early on January 1, 1899. Mann was promoted to Superintendent of Interlocking of the Boston project during its first year of service. He left Union Switch & Signal in 1903, and later became President of the Railway Signal Association in 1906.

Soon after, Union Switch & Signal built the world's largest interlocking system at the St. Louis Union Station in preparation for the 1904 World's Fair. One of the interlocking systems boasted 215 electro-pneumatic levers that operated 403 units, switches and signals. It was claimed that this system did the equivalent work of 696 levers of the London type interlocking. What began as a decade of financial burden and internal turmoil found a successful ending as the Company faced a new century with a growing reputation and the prospects of maintaining its competitive edge in the industry.

J. P. Coleman supervised the 1899 installation of this large interlocking machine at Massachusett's Boston Terminal South Station on the Boston & Maine Railroad. In 1983, when this photo was taken, the tower was still in service.

"The development of the power brake was one of the greatest events in the evolution of the art of land transportation, second only to the invention of the tubular boiler and of the Bessemer process for making steel; but the development of power signaling and interlocking was necessary to the full effects of the brake. One was the complement of the other."

– Henry G. Prout

The Union Switch & Signal Co.
Swissvale, Pa.
THE LANGUAGE OF FIXED SIGNALS

7

AUTOMATIC BLOCK SIGNALS
ONE ARM HIGH TWO POSITION

HOME SIGNAL   DISTANT SIGNAL

PROCEED   STOP   PROCEED   CAUTION

MEANINGS
HOME SIGNAL
PROCEED—Block is in condition for train to proceed.
STOP—Stop and wait prescribed time, then proceed with caution, expecting to find train in block, misplaced switch or broken rail.
DISTANT SIGNAL
PROCEED—Expect to find next Home Signal in Proceed position.
CAUTION—Prepare to stop at next Home Signal.

1907		J U L Y		1907		
Sun.	Mon.	Tue.	Wed.	Thu.	Fri.	Sat.
	1	2	3	4	5	6
7	8	9	10	11	12	13
14	15	16	17	18	19	20
21	22	23	24	25	26	27
28	29	30	31			

This 1907 calendar on a cardboard placard was most likely a leave-behind for customers. The highly decorative Union Switch & Signal logos within the columns at the bottom are a far cry from the patented logo to come.

5   A New Century Begins

It was a new century, full of promise and the hope of prosperity. The depression ended in 1897 and unbeknownst to the industry, the years before World War I would become the golden age of the railroad passenger train in America – the years before the automobile and airplane boom. Passenger rates were more rigidly regulated than freight, providing better bargains each year throughout the Progressive Era from the 1890s to 1920s. The growth rate of railroads nearly tripled between 1896 and 1916. Almost since the beginning of the railroad age, passenger traffic had been falling relative to freight, although there had been a time in the 1880s when freight rate wars made passenger business somewhat more profitable. Pullman excursions mirrored the prosperous times, soaring from five million in 1900 to 26 million in 1914.

George Westinghouse's private parlor, dining and sleeping car carried the industry magnate, his family, guests and clients in style.

Perception of service during this time improved over the early criticisms of the mid 1800s. According to *Enterprise Denied*, "Foreign travelers reported that American railroads were unmatched for convenience, comfort, speed and dependability (The safety of the trains was still debatable) …Strong competition…ensured punctuality, as did the much heavier locomotives used to pull passenger trains 'that weigh as much as an English freight train.'"

There is a disparity in reports as to whether or not route mileage increased after 1900. Some sources state that the total main track mileage rose from 193,000 to 258,000 by 1915, with "second track mileage increasing 150 percent from 12,000 to 29,000" in the same time frame. Other sources assert that although route mileage did not drastically increase, rigorous upgrades involved laying "thousands of miles of additional mainline tracks"… and "extensive line work that included changing grades, straightening lines and replacing bridges." By the end of 1900, the mainline from Jersey City, New Jersey, to Harrisburg, Pennsylvania, was four tracks wide. By 1910, the Santa Fe was double tracking at a rate of nearly 255 miles a year. In *The History of Pittsburgh*, in 1910, "Only 129 miles still remain to be four track between the Capital and Pittsburgh."

Alexander J. Cassatt, president of the Pennsylvania Railroad Company, said in his 1902 annual report that the increase in traffic had outstripped facilities at virtually every point on the system. In the early part of the decade, Pittsburgh faced constant improvements to keep pace with the 20 percent yearly increase

Alexander Cassatt

The shops at Swissvale were a noisy cacophony of friction drop hammers, shear punches, air compressors and Westinghouse gas engines.

in tonnage. In 1903, there was one point on the trunk line between Harrisburg and Pittsburgh where one could count 28 passenger and 140 freight trains whizzing by in 24 hours. Pittsburgh's total rail tonnage, not including freight in transit, was greater than the combined tonnage of New York, Boston and Chicago. In 1905, that amounted to 92 million tons. Seven railroads entered Pittsburgh and another two, The Lake Shore & Michigan Central Railway and the Erie Railroad, had arrangements with the Pittsburgh & Lake Erie Railroad to handle freight in and out of the city. Pittsburgh could accommodate a remarkable 664 passenger trains and 20,000 to 25,000 cars daily. All of this growth meant good business for manufacturers of rail safety equipment, and during this decade, inventions were prolific.

George Westinghouse had his men bustling with work. His new shops, covering 308,520 square feet, were described in the March 14, 1902, Railroad Gazette as "what is probably the largest establishment in the world for making signals and signal machinery." The shops were a noisy cacophony of friction drop hammers, shear punches, air compressors and Westinghouse 125-h.p. gas engines, all powered by natural gas from nearby wells. Of course, he owned those wells too. Westinghouse was the first man in Pittsburgh to drill for natural gas, and he did so on his own estate at Homewood.

It was another of his calculated risks that paid off. When drillers finally struck "gold," he invited his friends over and tested the quality of the gas by lighting an oil-soaked rag and tossing it into the geyser. The flame burst over 100 feet into the air, and people came from all over the area to see the phenomenal sight. George Westinghouse invested in a web of pipelines throughout the city, and in 1884, he organized the Philadelphia Company to supply natural gas for domestic and industrial purposes in Pittsburgh.

Shop employees at Union Switch & Signal, circa 1900.

One of the new products Union Switch & Signal offered the railroads in 1902 was the first solid color red glass lens, developed by Nikolas Kopp, chief scientist for the Pittsburgh Lamp and Glass Company. Kopp developed a cadmium sulfide glass that was superior to the almost black and nearly opaque red glass previously used to manufacture signal roundels and lenses. The new development provided for significantly higher light penetration. (It was necessary to spread the old red glass in very thin layers over transparent glass in the desired shape. This method was known as "plating" or "flashing.")

From 1902 to 1909, Louis Henry Thullen came from Columbiana, Ohio, to work for Union Switch & Signal. His capabilities as an inventor were something to be reckoned with – he was considered by some to be the most prolific inventor in the company "when measured by the quantity and the commercial importance of his inventions." It all began when Thullen and another inventor, Fitzhugh Townsend, who worked for competitor General Railway Signal, independently came up with the same idea for a double-rail track circuit in late 1903 or early 1904.

Jacob Baker Struble

At the same time, another Union Switch & Signal inventor named Jacob Baker Struble (who had worked for the company since 1889), received a patent for a single-rail track circuit in 1903. In such a circuit, an electric locomotive draws DC power from an overhead power line and uses one rail to return the traction motor power (the power that propels the locomotive) back to the railroad's traction substation (a small power plant), to complete its circuit. The other rail, instead of being a long "ribbon" of electrical current, is divided into electrically isolated sections, each with its own electrical current. This allows one rail to carry the propulsion power and the other to carry the low voltage track circuit power for signaling. A small piece of electrical insulating material is spliced between the ends of these rail sections to keep them isolated. The AC track circuit signal is applied to these rail sections only, not to the rail carrying the powerful traction current. The same single-rail track circuits are still common today. They are simple, reliable and inexpensive.

In a double-rail track circuit, both running rails are used to return the traction power back to the substation. The rails carry the power that run the train. The question was: how do you get the weak signal on the rail (the AC track circuit for signaling devices) to run on the same rail as the high voltage DC traction current without the latter feeding back into the track circuit wires and destroying the track circuit equipment?

Where several others had patented unsuccessful designs, Thullen and Townsend independently discovered a method to overlay AC track circuits on the rails using insulated joints without disrupting the flow of DC traction current between the train and the traction substation. Both men designed an isolated, balanced AC track circuit with an electrically neutral mid-point connection at both ends. A new device, the "impedance bond," created this neutral connection that "drained off" traction current. The impedance bond is typically a low-profile unit about the size and shape of a large suitcase that is mounted in pairs near the insulated rail joints, partially buried in the rock ballast between the ties of the track. Connected to the rails and each other with heavy-duty cables, the bonds prevent the AC track circuit signal from mixing with the DC traction power current.

The inventors each submitted their patent applications so close to one another that an interference case developed at the U.S. Patent Office. The Patent Office awarded the invention to Thullen based on its "first to invent" basis, and the fact that he had shown more effort in "reducing his invention to practice." Litigation ensued between General Railway Signal and Union Switch & Signal until 1910, when both companies realized that a patent cross-licensing agreement was a more financially viable solution. This set the course for similar signaling business agreements among competitors in the United States.

While working at Union Switch & Signal, Thullen was additionally credited with patents for his pioneering work on the vane relay and the slow-release magnets (a type of electro-magnet), both of which are still in production today. The vane relay is used in track circuits that monitor electrified railroad tracks. It detects when a train occupies a track circuit and feeds this indication (as an electrical signal) to the controlling wayside system. The unit contains a small swing arm, or "vane," which reacts to two adjacent coils. When these coils are properly energized, relative to each other, the vane arm swings up and closes an electrical contact. This contact closes the track-occupancy indication circuit. If the coils are not properly energized, the vane arm remains safely in the dropped position by gravity.

Louis Henry Thullen was credited for his pioneering work on three 1906 inventions: the vane relay, the impedance bond, and the slow release magnet, while working at Union Switch & Signal.

The slow-release magnet is the forerunner and critical component of today's slow-release vital relays. It is designed to take a set period of time to fully de-energize when electrical current is removed. Standard relays, on the other hand, de-energize the instant current is removed.

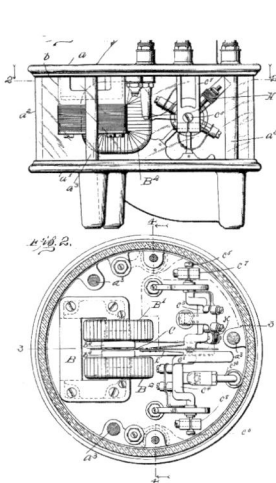

Vane Relay

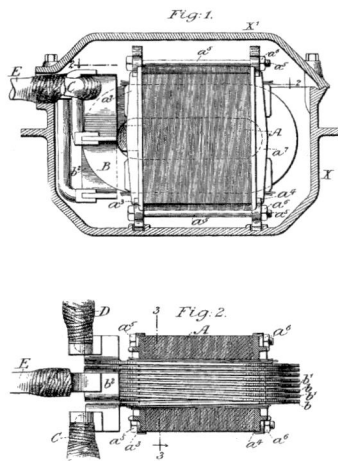

Impedance Bond

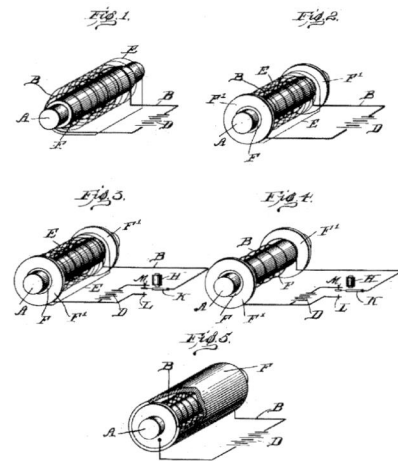

Slow Release Magnet

"...during the enormous development of the last four years, the railroads have found it very difficult to keep pace with the require-ments imposed upon them, and the so-called surplus earnings, as well as addi-tional capital, have been devoted to providing additional facilities... This work...must go on [and] during the next decade every single track railroad in the country will have to be double tracked, and provide enlarged terminal and other facilities."

—Edward H. Harriman to Theodore Roosevelt, 1904

In some vital relay systems, this delay is needed to properly time the closing or opening, as required, of another circuit.

These and other inven-tions fell under the daily opera-tions of the General Manager. In 1903, Colonel Henry Goslee Prout, past editor of the Rail-road Gazette, accepted those responsibilities as the Vice President and General Man-ager of Union Switch & Signal, and became general advisor to George Westinghouse in the latter's many interests. Prout recognized the demand for product was outgrowing their capabilities. In 1907, the Swiss-vale plant expanded, adding 34 acres of land and 157,000 square feet of shop space, in-cluding a foundry, forge build-ing and boiler house.

Union Switch & Signal Dwarf Semaphore used on the 2-foot gauge Sandy River Rangely Lakes Railroad in Maine, circa 1902, and inner workings of the dwarf semaphore.
Photo credit: Joanne L. Harris

One example of the sudden expansion and rebuilding efforts of the railroads was the Illinois Central. In the early 1900s, they faced the monumental task of replacing hundreds of wooden bridges and trestles as many of these were structurally deficient. (In fact, most early railroad efforts erected such structures hastily and on the cheap.) They would not bear up under the new, heavier locomotives. At the same time, they

In the early 1900s, the Illinois Central Railroad faced the monumental task of replacing hundreds of wooden bridges and trestles, like this Railroad bridge over the Ohio River in Cairo, Illinois.

initiated a major double tracking project on their trunk line from Chicago to New Orleans. By 1906, the railroad was well on its way toward the deserved title of "Main Line of Mid-America."

The last few months of 1906 marked the concurrence of the normal peak seasonal traffic and the highest level of economic activity that the United States (and, in fact, the Western world) had ever seen. Shippers had insufficient cars to haul the commodities. Eastern railroads, with barely enough fleet for their own customer demands, watched with their hands tied as the Western rail companies held on to their cars, refusing to return them promptly per American Railway Association rules. It was far cheaper to pay the "punitive" per diem rate to keep a car than to buy one for their fleet. In response, Theodore Roosevelt turned the ambitious Interstate Commerce Commission (ICC) Commissioner, Franklin K. Lane, loose on the railroads. Lane launched a series of crusading investigations of the railroad's inability to meet the nation's cargo transportation demands.

Franklin K. Lane

Another financial stall-out, this one called the "panic of 1907," occurred when the New York Stock Exchange closed 50 percent lower than its 1906 peak. A failed stock bid for the United Copper Company led to a series of bank runs that brought down the mighty Knickerbocker Trust Company, the third largest bank in New York City. According to *Enterprise Denied*, "If the Panic of 1907 had not happened, it might have had to be invented, so crippling was the lush prosperity which gripped the nation…the congestion on the railroads was a harbinger of things to come."

The panic of 1907 and the ensuing short-term, but cutting, recession translated into a severe drop in business travel and a sharp decline in the amount of freight being offered. Freight traffic fell by 7 percent from 1907 to 1908. Edward P. Ripley, president of the Santa Fe Railroad, had foreseen the railroad's problems just before the panic struck. Money was scarce, as investors had lost their confidence in the security and stability of railroad investments. The following year, he had to report to disappointed stockholders that for the first time in the history of the company that there were "no plans in hand for the construction of extensions during the year."

The American railroads' sudden growth spurt during 1906 affected much more than their purse strings. By 1907, with almost every mile

Customers flooded the banks on Wall Street during the bank panic of October 1907.

Twenty-five passengers were killed and 27 injured in a head-on collision between the Quebec to Boston express and a Boston & Maine freight train in Canaan, New Hampshire, on September 15, 1907.

of the most-used tracks under repair or reconstruction, scores of inexperienced, unskilled crews were hired to work on the right-of-ways during the time when a record volume of traffic was using the tracks. Due to the lack of trained labor, men were handed responsibilities beyond their abilities, resulting in a high accident rate for both passengers and employees. According to the ICC's annual report, passenger deaths surged from 359 in 1906 to 610 in 1907, and employees killed on the job jumped from 3,929 to 4,534. In addition, 1907 saw the highest number of injuries in the United States – an astounding 87,644 employees and 13,041 passengers, trespassers or innocent bystanders were injured on or by trains. A move toward new equipment and improved rights-of-way was rewarded by a steady downward trend from that point onward.

In fact, American railroads could never have handled the traffic volume in their terminals without technological advances in signaling. Central office train dispatching was widespread in the United States by 1900. Both automatic and remote switching and signaling equipment installed on the lines represented an aggressive development of the interlocking systems. By 1907, the railroads began to increase the number of automatic block signals on main lines. The Eastern trunk lines had installed the initial block systems some years prior, and now other railroad companies looked to the switch and signal industry to manufacture large numbers of these safety systems. The program fell under the sword of the Panic of 1907, in which the usual

McKee Rankin's Lost Opportunity

New York Press, Pittsburgh Gazette Times, August 28, 1909

McKee Rankin's favorite poem is "Opportunity," the sonnet that stands as the monument to John J. Ingalls. When Rankin was in the heyday of his success as an actor, a youth offered him an interest in an invention for $500. The young man was in sore need of the money; he had sublime faith in his invention, and he wanted that $500 to cover his patents. He once had met Rankin, and now turned to him as the only man he knew with more than his shirt to his back. Rankin wasted no time on the youth. The young man was not discouraged by the first visit, but came again the next night, and the next, and wound up by pestering the actor until orders were issued to keep him out of the theater.

Twenty years rolled around. Rankin happened to be in Chicago. Age was crowding him fast and the future looked black. A bellboy entered his room and said a friend wished to see him, but that he refused to give his name. "Anybody's welcome," said Rankin, "show him up." A big, broad-shouldered man with the mark of prosperity on him walked into the room. "Do you remember me, Mr. Rankin?" asked the caller. "I can't say that I do," replied Rankin. "Well, do you remember that crazy young fellow 20 years ago with his airbrake invention and the way he was put for $500?" Rankin sat up straight. "Why, yes, I remember; what about him?" "Oh, nothing, only that I'm that young fellow, and I just thought I'd drop in and say hello. I'll bet you don't even remember my name." "I've got to confess I don't," said Rankin. "Here's my card, then," said the visitor. Rankin took the card and read: "George Westinghouse."

McKee Rankin

The 1908 Signal Directory depicts the range of railroad crossing signs that were available from an assortment of suppliers.

a. Union Switch & Signal crossing bell, relay box and battery case with a wooden post

b. Bryant Zinc Company crossing bell, crisscross sign and relay box on an iron post

c. Union Switch & Signal crossing bell, round crossing sign, relay box and batter case affixed to an iron post

d. Bryant Zinc Company crossing bell, open sign and relay box on an iron post

e. Bryant Zinc Company bell, rectangular sign and relay box on an iron post

f. Railroad Supply Company crossing bell, crisscross sign and relay box on a channel iron post

a.

b.

c.

d.

e.

f.

—Superior Privy Councilor
W. Hoff and Privy Councilor
F. Schwabach, 1906

suspect – a shortage of capital – sent many projects reeling into oblivion. By 1912, less than 10 percent (20,000) of all track miles were protected by automatic block signals.

In 1907, however, the Pennsylvania and the Baltimore & Ohio Railroad Companies awarded Union Switch & Signal the entire contract for the interlocking and signaling of Union Station under construction in Washington, D. C. The installation was electro-pneumatic, similar in characteristics to the system in the Pennsylvania Railroad's Union Station railway yards in Pittsburgh. The Washington installation was, at the time, the largest and most complicated signaling and interlocking installation that had ever been constructed in the world. It was their third "largest interlocking system in the world" project in one decade. Previously, the largest system was the St. Louis job for the 1904 World's Fair. Before that, the Boston Terminal South Station was their most notable success story.

Along with other suppliers, Union Switch & Signal felt the effects of the 1907 panic, reporting 1908 gross sales at $2,103,718.49, considerably less than half of the sales of 1907's reporting of $4,993,731. The profit was only $77,848.68 against $799,598 in 1907. Fortunately, the setback was brief. In 1909, the company showed a gross sales increase of $507,938, or 24.1 percent over 1908.

Winding down the decade, Union Switch & Signal installed a new gas power plant at Swissvale in 1908. In 1909, the company shortened its name to the "Union Switch & Signal Company." The same year, the current Union Switch & Signal "Pretzel" logo was designed, although it was not formally registered as a trademark until 1956.

"The Union Switch & Signal "pretzel" logo, designed in 1909, remains one of the most recognized trademarks in rail transportation. The logo is now retired from corporate use, but can still be seen today on metal casted products from the Batesburg manufacturing plant."

Built in 1916 by Baldwin Locomotive Works of Philadelphia, Pennsylvania, this Pennsylvania Railroad class L1 Mikado (2-8-2) locomotive ran during The Great War (WWI) until its boiler exploded due to low water.

Photo Credit: Joanne L. Harris © 2011

6   Turmoil, Progress and a Call to Arms

The 1910s brought sweeping change to Union Switch & Signal, both externally and internally. The government was now investigating the technology and economics of train control, the Interstate Commerce Commission (ICC) reported high incidences of deaths and injuries caused by rail accidents, and demand for automatic block signaling was booming. Despite the Union Switch & Signal – General Railway Signal patent cross-licensing agreement of 1910, competition from the three major players – General Railway Signal, Hall Signal and Federal Signal– was a growing threat to Union Switch & Signal's market position.

In the first quarter of 1911, the ICC reported 2,124 killed and 16,430 injured, including employees, passengers on trains or boarding/deboarding, travelers at highway crossings, and trespassers. This figure did not include electric line statistics. One collision involved a passenger train rear-ending another passenger train that was standing at a station. The train had passed two automatic signals set against it, and it struck the standing train still running at high speed, killing 6 passengers and injuring 46 passengers and employees. The 54-year-old engineman had had a clean record for 30 years. It was believed that he had been unconscious for ten minutes immediately before the collision, due to a mild attack of epilepsy. The novice fireman was attending to the firebox and neglected the rule to verbally inform the engineman of each approaching block signal. He was also held negligent for not observing the excessive speed of the train as it approached the station.

A Union Switch & Signal wig wag signal (the DW Flagman) swung in a pendulum fashion to warn street vehicles that a train was coming. The swinging sign warned people to "Look and Listen" at the Delaware, Lackawanna and Western Railroad crossing at Chenango Forks, New York, circa 1918. An active DW Flagman was still in use in Joplin, Missouri, in 2006.

In 1911, an American Electric Railway committee concluded that ten of the 50 railways they researched were using automatic signals with continuous track circuits. Union Switch & Signal led the pack with seven of those installations. On the railroad side of business, Union Switch & Signal won a contract for the block and interlocking signaling on the New York, Westchester & Boston Railroad, comprising 32 automatic signals on a six-mile section of four tracks; and another 32 automatic signals on a two sections of double track. The interlockings were power operated, all-electric Union style "F" type, operating at 110 volts on storage battery. The popular Union style "B" mechanism operated the center-pivoted two-position home and distant

The Hepburn Act of 1906

Whenever a government passes down regulation to one industry, it invariably creates an economic trickle-down effect for all other industries that supplies goods or services to it.

The Hepburn Act of 1906 brought express companies under ICC regulation, which effectively froze rates until World War I began. Express companies split profits with the railroad to ship their packages, and business was good for everyone, except that the railroad also had to carry the mail. The mail rates were based on weight, which was only assessed every four years, such as in 1911.

William Hepburn

When the Postmaster General inaugurated parcel post service in 1913, the railroads were required to carry it at the same rate as the regular mail. Whether or not Congress realized the potential impact of this act on the railroads and the existing express services is not known, but the fallout was extensive. Mail tonnage soared 25 percent; express volume plummeted.

The Adams Express Company and the United States Express Company were wiped out, and American Express was nearly wiped out. In "response," the ICC reduced the rates by 20 percent in 1914 on what paltry express business was left.

The revenue loss for railroad and express companies was estimated at $50 million. Parcels previously shipped in express cars, where the railroads were paid to transport them, were now riding free of charge in U.S. mail cars. he railroads approached Congress with their position and Congress agreed to establish volume limits.

Railway mail trains were required to haul excessive loads of parcels at the same price as mail, affecting both railroad and express service profits.

blades. They also furnished automatic block signals and three-position, upper quadrant semaphores on about 40 miles of the Erie Railroad's lines, and they won the contract to erect mechanical interlocking systems at locations in Polk and Burbank, Ohio.

The General Electric Company in New York sold its nascent signaling business, including all of its patents relating to railway signaling and the departments' stock and manufacturing equipment, to Union Switch & Signal the same year. This allowed the latter company to offer complete signal installations. General Electric had entered the signaling business in 1905 after acquiring the signaling-related patents of Reinhold Herman, Max Hanna, Lawrence Hawkins and Frederick Bedell. One of their prized engineers, Fred B. Corey, a Cornell University graduate, left to work in the Union Switch & Signal plant from 1911 to 1914 as Engineer of Test and Inspections – what we would today call a Quality Assurance Manager. Corey amassed a collection of over 60 patents in his career, many of which were assigned to General Electric or Union Switch & Signal. At least 12 patents were directly related to railway signaling or train control and several more were for vehicle-borne traction power and braking systems.

Following the acquisition of the General Electric Company's railway signaling business, Union Switch & Signal entered into a trade agreement with the Australian affiliate of General Electric to exclusively represent them in Australia and New Zealand from 1912 to 1926. Similar agreements were made in 1913

Union Switch & Signal supplied automatic block signals and three-position, upper quadrant semaphore signals on 40 miles of the Washington, Baltimore & Annapolis Electric Railroad circa 1914.

Union Switch & Signal developed an apprenticeship program in 1913 to compete with larger companies to entice the most talented engineers to work at Swissvale. Here, a later group of recruits.

with General Electric's South African and South American affiliates. These international agreements rounded out the Union Switch & Signal's growing market leadership at home.

In the United States, one of the most successful projects that Union Switch & Signal accomplished at this time was the Interurban automatic block signaling on the Washington, Baltimore & Annapolis Electric Railroad between Naval Academy junction and Annapolis, Maryland. The Railroad connected the three cities with a high grade, high speed, 1,200-volt DC trolley line. Union Switch & Signal outfitted a double track road between Baltimore and Washington, a single-track line from Annapolis Junction where it connected with the B&O Railroad, and the 13.2-mile-long portion from Naval Academy Junction to Annapolis with automatic block signals prior to the Democratic Convention of 1912. The entire automatic signaling system was installed in record time – *the signals were in service 30 days after the first shipment of material and 49 days after the award of the contract.*

It was vital to attract the brightest talent to overcome the challenges of so many great projects. To compete against high-capital corporations who could offer higher wages, Union Switch & Signal developed a two-year apprenticeship program in 1913 to entice young, college-educated men to join the company. Apprentices learned the signaling appliances, installation, and design engineering with wages of $65 per month for the first year, and $75 per month for the second. The Company even paid the entrance fee and one year's dues to the Westinghouse Club in Wilkinsburg, located two miles west of Swissvale. Most Union Switch & Signal Company employees lived in Wilkinsburg, considered at the time to be "one of the most desirable residential sections of the Pittsburgh district." The town boasted a population of 20,000 and offered 20 churches of 11 denominations within its borders. The Club gave employees in any of the Westinghouse industries "opportunities to meet the engineers and older men of the various companies in a social way…enabling a man to build up a wide and valuable acquaintance." Or, as we call it today, "networking opportunities."

The club's social life included dances and entertainment, lectures, a fully equipped gym and outdoor sports facilities – and of course, a subscription to the Electric Journal. Apprentices belonged to the Signal Club, which held semi-monthly meetings in the Westinghouse Club rooms to study the art of signaling.

Union Switch & Signal commanded a large presence at the Chicago Exhibitions. Featured here at the Coliseum in 1916.

The Company stated, "…we will use our best endeavors to retain them within our own organization, or, if that is not possible, we will try to secure them employment with the various railroads." This strategy offered a win-win solution. If they placed them with a customer, the railroad gained engineering personnel who knew the new signal equipment and the supplier earned an inside connection.

While Westinghouse remained President of the company, his failing health kept him inactive in the |day-to-day affairs. Thus Colonel Prout became the active head in addition to his duties as Vice President and General Manager. One of his early decisions was to take out foreign patents in Britain, France, Germany, Australia and New Zealand in the name of Union Switch & Signal on certain inventions, primarily the Follett Type F interlocking and the Taylor Style T2 alternating current signal. Under his ten years of management, according to the Railway Age Gazette in 1914, the "company has grown from a corporation with $2,500,000 stock and net income of about $300,000, to a corporation with $10,000,000 stock authorized and close to $7,000,000 issued, and net income for 1913 of $1,618,000."

Sidney G. Johnson was appointed Vice President of Sales and Engineering in 1914. He came to America from Eccles, Lancashire, England, at the age of 12. He landed in Swissvale where his father, Henry Johnson, was Works Manager at Union Switch & Signal. (Henry Johnson was the brother of Charles R. Johnson, the Company's General Manager from 1886 to 1888.) Sidney Johnson worked at the Johnson Railroad Signal Company in New Jersey during school vacations. At 21, he worked briefly as a field workman for Union Switch & Signal before following his father to the newly organized Standard Railway Signal Company in 1896. Sidney Johnson hired on as a signal engineer, but after his father left, Sidney returned to Union Switch & Signal in 1899. He worked his way up from Engineer of Construction for the Eastern District to Eastern Manager in charge of sales and construction, and later became General Sales Manager, based in New York. Both of these men would soon become intricately involved in the 1914 Wood's scandal.

Arthur Humphrey reflected on George Westinghouse at an address he gave at the Annual Banquet of the Westinghouse Veteran Employees' Association in 1915. In this photo are the 1914 Annual Banquet honorees.

Arthur L. Humphrey

"[Westinghouse] never for a moment lost interest in an individual working man. He knew when to say a kind word, he knew that a little encouragement is the oil that smoothes the machinery of success, the caress which sets wagging the tail of the world...I was especially impressed by his thoughtfulness for his older employees."

— A. L. Humphrey

By 1914, all of the Midwestern railroads were in grave financial trouble. Gross revenues were not rising as rapidly as on the trunk lines. On the Chicago & Alton (which prior to 1898 was a conservatively financed and very prosperous railroad), operating expenses rose to $12.3 million in 1914, just under their gross revenues of $13.7 million. No railroad could exist under such conditions.

Westinghouse died of chronic myocarditis (heart disease) in New York on March 12, 1914. He was 68 years old. He had limited his corporate activities as the President of 30 corporations worldwide for three years due to his failing health. A. L. Humphrey, as Westinghouse Air Brake Company Vice President and General Manager, sent a condolence telegram to Mrs. Westinghouse and wired all the Westinghouse companies and their worldwide offices to shut down their works for the day. This brief shutdown affected about 25,000 men in the Pittsburgh district alone. Eight of the oldest employees from Westinghouse Air Brake were approached and volunteered to serve as the active pallbearers in New York.

Humphrey later reflected on Westinghouse at an address he gave at the 13th Annual Banquet of the Westinghouse Veteran Employees' Association on October 28, 1915, in Wilmerding. "He was always impetuous and in a hurry, but I never knew a man in my life who could put other people so much at ease," said Humphrey.

"He never for a moment lost interest in an individual working man. He knew when to say a kind word, he knew that a little encouragement is the oil that smoothes the machinery of success, the caress which sets wagging the tail of the world…I was especially impressed by his thoughtfulness for his older employees."

It was a mere three months after Westinghouse had been laid to rest when, on June 12, 1914, the Allegheny Congenial Industrial Union (ACIU) organized a strike against the Westinghouse companies in Swissvale and the Turtle Creek Valley. Between 1,100 and 1,400 Union Switch & Signal employees, many of them women working as coil winders, walked out at lunchtime to picket with about 2,000 other Westinghouse company strikers. Women played a vital role in the strike. One in particular was Bridget Kenny, whom Westinghouse had fired in 1913 for selling union benefit tickets on company grounds. Kenny, dubbed the "Joan de Arc" of the strike by the Pittsburgh Leader, organized marches and recruited workers into the ACIU. Union Switch & Signal General Manager Colonel Prout told the press, "The strikers' action seemed to have been spontaneous, and came without intimation to us of their intentions." Later that day, the strikers stated their demands – an eight-hour day, reinstatement of discharged union men, permission for workmen to elect grievance committees, and higher overtime and holiday rates. Prout announced that the works would shut down until Monday, at which time the company expected employees to return to work – and most of them did return.

The strike was one of many to follow in the United States. The country was in an economic transition from war to peace. Unions were defeated in the 1919 steel strike, the 1920 men's clothing industry strike,

Strikers paraded down Edgewood Avenue in Pittsburgh during the 1914 ACIU strike. (inset) One of the first visual references to the company's nickname – "The Switch"

and the 1919 and 1922 coal-mining strikes. The breakup of the Socialist Party from 1918 to 1919 crushed a movement that had engendered union leadership for electrical workers.

The internal turmoil continued. In late June or early July of 1914, in the presence of Walter Uptegraff, Sidney Johnson made the career-breaking mistake of suggesting to the newly appointed President, Henry G. Prout, that a $5,000 bribe would purchase Robert C. Wood's tie-breaker vote to win a bid on the Kansas City, Clay County and St. Joseph Railroad's Centre St. Loop subway system project. Wood, then President of the Northwestern Construction Company, was soon to become Commissioner of the Public Service Commission. Uptegraff, vice president in charge of finance, called a special Board of Directors meeting on July 15, 1914, and five of the nine directors met with Prout and Johnson. The directors knew of the dealing and had already planned to oust the two "bad apples," but held a meeting with them anyway. After hearing them out, the directors demanded their immediate resignations for suggesting that the money should be paid. General Railway Signal president Wilmer W. Salmon immediately hired Johnson, who remained with that company as the Vice President of Sales and as a Director for the next 37 years.

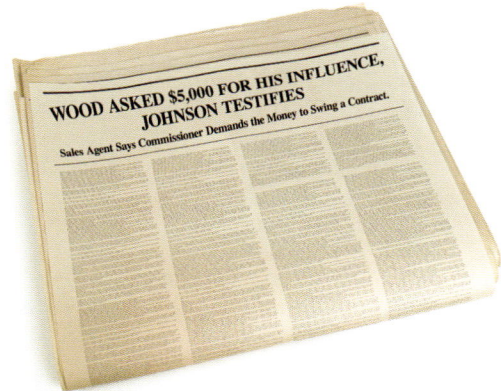

The scandal over a Public Service Commissioner's bribe set the wheels in motion. Regardless of actual blame, Prout was out.

Walter Uptegraff succeeded Prout as President and J. S. Hobson succeeded Johnson as General Sales Manager. Uptegraff later testified before the Thompson legislative committee that his company refused to give the money to Wood. Throughout the investigation, Wood defended the deal, claiming that it was made six months prior to his Commissioner appointment in February 1914. In reality, he was paid 13 days prior to his appointment. William C. Banks, who became manager of the Northwestern Construction Company after Wood made Commissioner, also testified that his business did very well after Wood was in office – and that the business had accordingly fallen off after the investigation started.

The internal conflicts of the Company resolved, Uptegraff and his colleagues turned their attention to the myriad of patent actions that 1916 brought between the four major signaling companies. Union Switch & Signal had filed suit for patent infringement against the Hall Signal Company, the Federal Signal Company and the New York Municipal Railways Corporation, the latter of which was a customer of both Hall Signal and Federal Signal. At the same time, Union Switch & Signal was defending one of its customers, the Western Maryland Railroad Company, against an infringement suit that General Railway Signal filed. These suits, along with a number of "bitterly contested interferences" in the Patent Office, were costly to both the companies involved as well as to their respective engineering staffs. According to Malcolm Farnsworth, patent attorney for Union Switch & Signal, the engineers spent "a large portion of their time in the preparation for and participation as witnesses in interference actions and patent litigation." The time had come for the four companies to find a solution.

Malcolm Mallory Farnsworth

A proposal was made for a patent interlicensing agreement in the field of railroad signaling, in a manner that each company would be licensed under the signaling patents of the other companies and would

Walter Denny Uptegraff

The Family Advisor

It is a significant achievement to become responsible for a growing corporation's financial department, but few men's stories, if any, have ever included an appointment outside the workplace to become a magnate's personal family accountant, private secretary and overall personal confidant.

For Walter Denny Uptegraff, a "home-grown" boy who grew up in Allegheny City (the same low-rent district where Robert Pitcairn and Andrew Carnegie were raised) such thoughts of a prestigious career were probably mythical at best. Born in 1865, he attended school until 1880, when he was 15. That year he found employment at the Westinghouse Air Brake as Secretary Howard Sprague's assistant. His humble beginnings as an office boy and stenographer were soon rewarded.

George Westinghouse trusted Walter Uptegraff like no other, handing over the keys to his personal secrets and correspondence for his many enterprises, leaning on him – when Uptegraff was only 31 years old – for financial advice and relying on him to act as the voice of the company in press matters. All this responsibility on the shoulders of a man with a ninth-grade education, and at some point in his career, a course of legal study, must have at some point seemed a bit overwhelming. Or perhaps not. After Westinghouse gave him Power of Attorney over his personal financial matters, Westinghouse went bankrupt twice in his last 25 years, yet he left an estate valued at $50 million in 1914. While some may consider bankruptcy to be a sign of a poor financial advisor, it is important to remember that the history of the headstrong Westinghouse revealed a man who always seemed more concerned about moving forward with costly experimentation than with the financial consequences – be it positive or negative.

Uptegraff was the closest and most influential of the many long-time associates of George Westinghouse, including Robert Pitcairn, brother Henry Westinghouse, Charles Terry and Paul Cravath. He moved to Wilkinsburg around 1900, where most Union Switch & Signal employees resided, and later moved into an apartment at Pittsburgh's Bellfield Dwellings.

Walter Uptegraff was elected a Union Switch & Signal director in March 1913. When Westinghouse passed away in April of 1914, Uptegraff served as one of three executors of his estate. He served briefly as Vice-President of Finance beginning in March 1914 after Colonel H. G. Prout succeeded the deceased Westinghouse as company President. Uptegraff became President pro-tem on July 15, 1914, after dismissing Henry Prout and Sidney Johnson in connection with the Wood's Affair. After the merger of Union Switch & Signal and Westinghouse Air Brake, Arthur L. Humphrey became President of Union Switch & Signal in 1917. Uptegraff moved into the position of Chairman of the Board, where he remained until his death in 1929.

Over the years, Uptegraff also became a Director for the Westinghouse Air Brake, the Westinghouse Machine Company, and Treasurer and Director of the Westinghouse Air Spring Company. He held one U.S. patent for a current regulator for the Nernst "Glower" lamp, which Westinghouse himself signed as witness on the patent application in 1909.

Walter Uptegraff was also President and Director of the Pittsburgh Wall Paper Company and the Defiance Paper Company; president, assistant secretary and director of the Excess Indicator Company; and treasurer, secretary and director of the East Pittsburgh Improvement Company.

At the Pittsburgh Wall Paper Company, Uptegraff, his brother James and several others owned and ran three wallpaper-related businesses, including a large paper mill in Niagara Falls that produced pre-pasted, do-it-yourself wallpaper. After his brother Thomas died in 1920, Walter moved to Niagara Falls.

Uptegraff belonged to the elite social clubs in Pittsburgh – the Duquesne Club, the Pittsburgh Country Club and the Pittsburgh Athletic Association. It was written that he had a "capacity for winning and holding friends," and a "straightforward disposition and the ability for prompt decision and unhesitating action." He enjoyed entertaining friends in his family home, along with his wife Annie Gaylor Marshall (whose mother was a cousin of Andrew Carnegie) and their eight children.

Walter Uptegraff served Westinghouse-affiliated companies for nearly 50 years before his death on February 17, 1929, one day prior to his 64th birthday.

in turn license the other companies under its signaling patents. The field of signaling now encompassed five active classes of business – two in block signaling and three in interlocking systems. Each of the four companies was engaged in some of the fields, but only Union Switch & Signal was engaged in all of them. Because of the disparity of patent contributions, it was clear that each company could not be expected to become licensed or to license the others on an equal basis. Thus, the agreements had to be examined based on profits that each patent had made over the previous five years compared with the total value of business done by the four companies during that timeframe. All was going well until the Hall Company withdrew from the negotiations after demanding more favorable terms than the other companies deemed appropriate. This resulted in a complicated, but realistic, three-party agreement dated March 20, 1916, between the Union Switch & Signal, General Railway Signal and the Federal Signal companies.

Less than two months later, Hall Signal changed its position again and requested to become a part of the agreement. On May 4, all parties agreed to the amendment and a Patent Bureau was set up with attorney George Cruse, to handle all issues regarding the agreements until Cruse's death in 1928. In 1917, the companies entered into a second agreement that established a committee comprising one representative of each company to investigate manufacturing costs for each company; conduct a publicity campaign to increase system sales; collect data of design, development and manufacture of signaling systems and apparatus; and perform data intelligence gathering of other patents and inventions relating to railway signaling.

In 1915, Union Switch & Signal entered into an agreement with Westinghouse Air Brake, McKenzie & Holland, Ltd.; Consolidated Signal Company and The Westinghouse Power Signal Company, in what was termed the "5-party agreement." Union Switch & Signal acquired a one-third interest in the Westinghouse Power Signal Company as a result. The agreement called for Union Switch & Signal to have exclusive rights to the territory of the American continent north of the Panama Canal, Cuba and Japan. The Westinghouse Power Signal Company had the British Empire (except Canada), British Protectorates and the Continents of Europe and Africa.

Union Switch & Signal again faced a significant internal change at its corporate office in 1917. A plan for its merger into the Westinghouse Air Brake was formally announced on January 12. A meeting of the directors of the Switch Company was later held to reorganize and give Air Brake officers representation on their board. Uptegraff was chosen as Chairman of the Board; A. L. Humphrey, first vice-president and general manager of Westinghouse Air Brake, became President; John F. Miller, president of Westinghouse Air Brake, was made Vice-President;

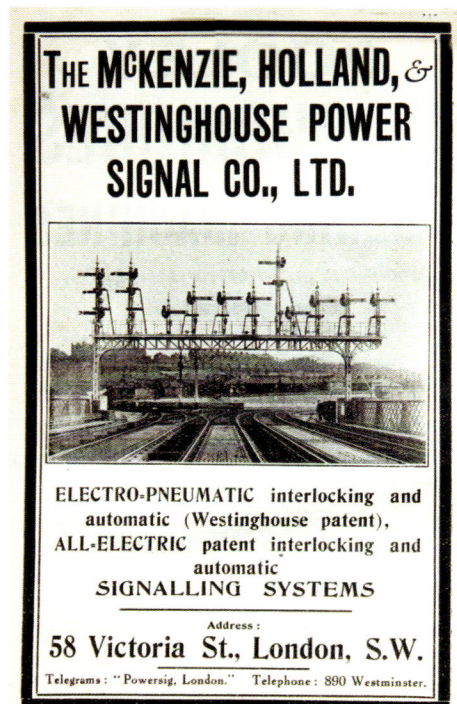

Union Switch & Signal branched out into other countries, including this London, England, firm in 1915.

A train using Westinghouse's air brakes screeches to a dramatic halt. Painting by Roy Hilton, an art instructor at the Carnegie Institute of Technology (now Carnegie Mellon University), in Pittsburgh.

Union Switch & Signal's main machine shop had been destroyed by fire on February 10, 1917, with a large part of its machinery and a large amount of finished or in process-product material. The loss was recorded as "one of the big losses of the year," calculated at $4 million.

T. W. Siemon, erstwhile vice-president of Union Switch & Signal, became Vice-President and Treasurer; G. A. Blackmore, former manager of sales of the Signal Company became Vice-President in charge of sales; and T. S. Grubbs, past secretary of the Signal Company, was voted acting Vice-President and Secretary. Westinghouse Air Brake directors called a special stockholders meeting on March 15 to ratify the merger and to approve an increase in capital from $20 million to $30 million to finance the transaction. The same year, Union Switch & Signal began filing for patent protection in a number of countries. The majority of the patents – about 30 – were filed in Canada.

Humphrey later shared with the employees of Westinghouse Air Brake and Union Switch & Signal, during a speaking engagement in 1919, that this merger fulfilled Westinghouse's vision. In his address, he said, "I am very proud to be able to offer you on behalf of the men of Union Switch & Signal a hearty welcome in these shops. It is an occasion which brings to actual realization a cherished dream long entertained by our former chief, the great founder of our companies, who often told me during his lifetime, that he hoped some day to see the Air Brake Company and Union Switch & Signal form one organization."

A terrible fire destroyed the main machine shop, but it could not destroy the resolve of Arthur Humphrey and his employees as they set forth to manufacture the war materiel in support of "the Great War" just seven months later.

World War I broke out in Europe after Archduke Franz Ferdinand of Austria, heir to the throne of Austria, was assassinated on June 28, 1914. It was not until 1917, when the most powerful countries in the world were at war, that Union Switch & Signal was called upon by its government to assist in manufacturing war material. Union Switch & Signal 's main machine shop had been destroyed by fire on February 10, 1917, along with a large part of its machinery and a large amount of finished or in process-product material. The loss was recorded as "one of the big losses of the year," calculated at $4 million. Any available space in existing buildings was quickly reassigned to accommodate shop machinery, and the Company began construction on two five-story buildings. The company was still in the midst of this rebuilding when in September, the U. S. government contracted them to produce 2,500 Le Rhône engines for the country's infant Air Command.

The American-built Thomas-Morse S-4 Scout biplane was an advanced trainer flown during World War I.

The 80-horsepower, nine-cylinder engine, designed by Louis Verdet in France, was widely licensed in Germany (by Oberursel, whose Le Rhone engine copies received an "UR" prefix), Austria, Britain and Sweden. In the United States, it was licensed to Union Switch & Signal for use on the Tommy Morse Scout airplane built by Morse Chain Company of Ithaca, New York. Union Switch & Signal received a sample engine, drawings and material specifications in September, with production completion scheduled for July 1, 1918. They completed the two new buildings in December 1917. Time was short. Employees on the project were promised a 10 percent bonus, contingent on the Company turning a profit on the contract. The initial cost-plus estimate of $5,200 per engine was higher than actual production costs, and in mid-1918, the government renegotiated the unit price to $3,250.

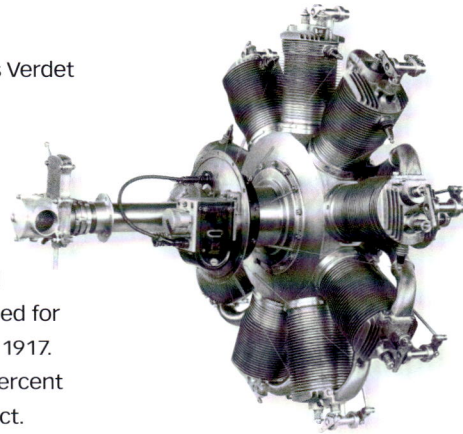

The United States Army touted Union Switch & Signal's 80-horsepower, nine-cylinder Le Rhône engines as "the best constructed rotary engines which have ever been built."

The Le Rhône engine was unique in that its cylinders and crankcase revolved around a stationary crankshaft. Rotary engines of this type were noted for high reliability. Castor oil, of all things, was used as a lubricant, mixed with fuel inside the crankcase. In July 1918, there were reports of a castor oil shortage that the rotaries required. A new Voltol-based lubricant was substituted, but it was blamed for a rash of engine failures on German Fokker E.V. aircraft that were outfitted with the Oberursel U.R. II rotary engine.

Union Switch & Signal received an increased contract in early 1918 for a large order of non-flight engines for training purposes and spare parts, along with orders for raw and finished forgings to support other Le Rhône producers. M. Guillot, a design engineer for the Le Rhône factory who came from France to guide the American effort, was quoted in an Army Report dated January 1919, as saying, "These are the best constructed rotary engines which have ever been built."

According to a sales data sheet, Union Switch & Signal manufactured a wide array of war munitions and aircraft products for the World War I effort. This order included 4.5-inch shells, 4-inch shells shrapnel, base plates, rifle parts, 2,500 80-horsepower aircraft "flight" engines, 85 aircraft "non flight" engines, 2,500

Arthur Humphrey (at microphone) gave one of the world's first radio addresses on KDKA in 1922.

sets of spare parts, and 1,000 sets of aircraft forgings for roughly $13.5 million. About 2,300 engines were completed before the government issued a stop work order after the November 1918 armistice.

The Big War over, foreign patents were filed in France and Italy in the name of Compagnie Generale de Signalisation, the French affiliate owned jointly by the Westinghouse Power Signal Company and by the Compagnie des Freins Westinghouse.

In 1916, Dr. Frank Conrad, assistant Chief Engineer at the Westinghouse Electric Company, created the world's first broadcasting station in his second-story garage in Wilkinsburg. The first report that he broadcasted at "8XK" (the station's license) was the election returns from the Harding vs. Cox United States Presidential race. Of course, when Westinghouse Electric saw the clamor for radio receivers and the advent of radio advertising, they began to manufacture and sell amateur wireless sets so the public could "tune in." By October 1920, Westinghouse Electric applied for the first ever radio license – and the letters "KDKA" were assigned. The second key event in the development of radio science came on January 2, 1921, when, for the first time in history, church services were broadcast – in this instance being Calvary Episcopal Church in Pittsburgh. Arthur L. Humphrey, President of Westinghouse Air Brake, and a member of the Episcopalian church, gave one of the world's first address over the radio on KDKA in 1922.

Union Switch & Signal, with a New York sales office and district offices in Chicago, Atlanta and San Francisco, began offering a Style "T" upper quadrant Electric Semaphore Signal in 1910, the Model 15A DC Highway crossing bell in 1917, and the Style "M" Electric Switch & Lock Movement in 1918. The Style DW

Three Aspect Highway Crossing Flagman was also first produced around 1918, but for some reason, it was never patented. The Flagman did, however, find some degree of popularity on the Erie, Boston and Maine, Delaware Lackawanna and Western, and Canadian Pacific Railroads.

Westinghouse Air Brake ended the decade by publishing a monthly employee publication in 1919, covering events "affecting Westinghouse Air Brake Company and its allied interests." On September 27, 1919, the company held a 50th anniversary Jubilee dinner at the William Penn Hotel in Pittsburgh. They invited all Westinghouse Air Brake and subsidiary companies' employees, including Union Switch & Signal, who had more than 21 years of service. The event included dinner for 832 employees, along with music and entertainment. Of the guests, 138 represented Union Switch & Signal, including J. P. Coleman, Jenns Schreuder, V. K. Spicer, ex-president Walter Uptegraff, and two future presidents of the company, George A. Blackmore and W. H. Cadwallader.

A local character impersonator, Luke Barnett, played a joke on the guests by disrupting the entertainment, insisting that Humphrey let him address the Board members immediately over a critical rail issue. The joke was a complete success as Master of Ceremonies A. L. Humphrey played along, much to the chagrin of the employees who thought the gag was real. Humphrey then gave a formal speech regarding all of the companies, including an acknowledgement of Union Switch & Signal: "The importance and value of our association with the Signal Company is generally understood," said Humphrey. "We all know what that organization has accomplished in the important field of railroad signaling and interlocking in the past and what it will do in the future, if the railways are again permitted to breathe the breath of life. In its particular line the industry is only second in importance to that of the air brake. We are mighty proud of its wonderful achievement in the production of airplane motors. A finer exhibition of all-around competency, both in engineering and manufacture, does not exist."

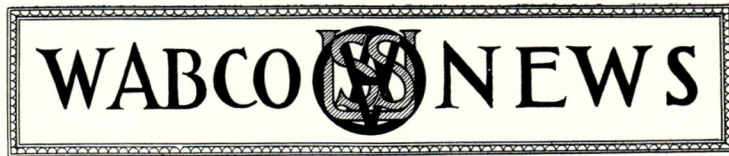

WABCO NEWS

VOL. I.   SEPTEMBER 1919,   No. 9.

My Acquaintance With the Air Brake Company—
Its Apparatus and Its Organization.

By A. L. HUMPHREY

When Westinghouse Air Brake brought Union Switch & Signal under its leadership, the company made efforts to show the solidarity of the two companies, as seen in this combined Flying W and Pretzel logo on the company newsletter.

An open box on the front of a Central Railroad locomotive in New Jersey displays the inner workings of Union Switch & Signal train control equipment.

7 Emerging Train Control Technology

American railroads were already in a financial crisis before World War I. The automobile and the electric interurban systems were, by 1916, absorbing large portions of what had been a steady, highly profitable local traffic. Heavy military demands compounded their fiscal problems when the government took control of the railroads in 1917 for 26 months, nearly running many railroads bankrupt and costing taxpayers $2 million a day. Freight car shortages skyrocketed to 158,000 *per day* as cars were blocked in at overcrowded East Coast ports, leaving shippers unable to transport food and bringing rail traffic to a near halt.

During this time, the Railroad Administration Automatic Train Control Committee began to investigate the need for train control apparatus. The Transportation Act of 1920 returned the railroads to private operation after the war ended. It also authorized the ICC to order any railroad to install automatic train stops or train control devices or other safety devices after appropriate investigation. In 1922, the ICC ordered 49 Railroads to install train control on complete passenger divisions. They issued a second mandate in 1924 that added an additional 31 new roads to the list.

Union Switch & Signal had developed a ramp type, speed control system in 1913, which was tested on several railroads, but when clearance difficulties indicated that the ramp type could not work properly under all weather conditions, the design was scrapped. Cornell graduate L. V. Lewis came to Union Switch & Signal to work on the development of printing telegraphs systems – a short-lived venture that ended with the Swissvale fire of 1917. He then developed an inductive system using General Electric's high vacuum "Pliotron" electronic tube. No one had previously used vacuum tube technology outside of the communication industry (i.e., telephones, telegraphs and radio), but Lewis discovered that vacuum tubes could amplify weak currents picked up by induction from the signal current in the rail. A complete laboratory demonstration in 1918, using track and loop circuits to provide three indications on the locomotive, proved that the vacuum tube could provide continuous inductive train control and cab signaling. This system, called the "Union" Automatic Continuous Train Control system, was demonstrated to ICC representatives in 1918, and later offered as two- or three-speed versions to railroad officials. This was an early form of today's continuous cab signaling, in which the wayside gives vital instructions to the train in a continuous, rather than intermittent, fashion.

L. V. Lewis

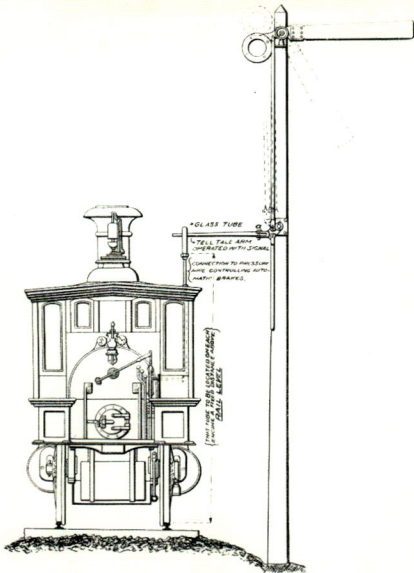

Vogt Automatic Train Stop Reproduced from The Union Switch & Signal Co. Reference Book of 1889.

Master Mechanic Axel Vogt of the Pennsylvania Railroad developed an automatic train stop that connected a glass tube on top of the locomotive cab to the automatic brakes. An extended lever on the semaphore pole would break the tube if the train did not stop for a distant "on."

In 1923, the Pennsylvania Railroad tested and put the "Union" three-speed train control system into service on 43.5 miles of single track and 3.4 miles of double track between Lewistown and Sunbury, Pennsylvania, marking the first time that cab signaling was used instead of wayside signals. The cab signal consisted of a miniature light signal that indicated what action the engineman must take according to upcoming track conditions. A warning whistle sounded if the signal changed to a more restrictive indication; it was silenced when the engineman flipped the "acknowledge" switch.

At the same time the Automatic Continuous Train Control system was developed, Dr. Lars O. Grondahl left his professorship at Pittsburgh's Carnegie Institute of Technology in 1920 to work for Union Switch & Signal as the head of research and engineering. His first project revealed that AC power could be "rectified" – made to flow in one direction only – using one side of a copper plate. This provided the foundation for a contactless switch, which became known as the copper

Dr. Lars O. Grondahl

oxide rectifier. The product was patented and introduced in 1927. The copper oxide rectifier was a major breakthrough in electronics history and the further development of signaling systems. It was used as a battery charger for DC track circuits, circumventing the need to rely on commercially supplied DC power. More reliable and readily available commercial AC power could now be used to supply track circuit power via the rectifier. This first electronic diode laid the foundation for the control diode, better known as the transistor, which in turn, provided the groundwork technology in the development of microchips. (Union Switch & Signal later donated Dr. Grondahl's laboratory notebook and his "breadboard" rectifier to the Smithsonian Institute in Washington, D.C.)

In the 1920s, railroad traffic studies showed that 75 percent of a freight car's life was wasted in terminal yards. Westinghouse had introduced remotely controlled electro-pneumatic switch machines in a flat switching yard at Altoona, Pennsylvania in 1891, eliminating hazardous manual work associated with rail yards. In 1923, Indiana Harbor Belt Railroad's Vice President George Hannauer, and master car builder Earl Wilcox developed a set of stationary rail brakes using air cylinders and pistons that squeezed the car wheels as they passed. The following year, the first installation of 51 "car retarders" produced 84 percent increase in yard capacity in the first year. The retarders significantly reduced the amount of time a car spent in the yard, and avoided collisions and other types of accidents. Within two years, in 1926, J. P. Coleman of Union Switch & Signal supervised the first installation of an electro-pneumatic version of the car retarder for hump yard switching service at the Markham Yards on the Illinois Central, which also was a tremendous success.

In the midst of all these new developments, the businesses of Union Switch & Signal and Westinghouse Air Brake expanded further on an international level. When the Westinghouse Brake Company and Consolidated Signal Company merged under the name Westinghouse Brake and Saxby Signal Company, Ltd.

Method of installing car retarders at the Portsmouth, Ohio, yards.

The car retarder squeezes against the car wheels to stop the car from moving at Markham Yard in Home-wood, south of Chicago.

in 1920, Union Switch & Signal became a stockholder in the newly formed company. This merger "associated" The Westinghouse Brake Company, Ltd.; McKenzie & Holland, Ltd.; Saxby & Farmer, Ltd.; and The McKenzie Holland & Westinghouse Power Signal Company, Ltd.

Union Switch & Signal conducted a series of international business deals, beginning in 1921 with patent protection in Japan on several system developments. They sent engineering representative S. E. Gillespie to assist the conglomerate Mitsui in selling Union Switch & Signal equipment, and he remained there until 1926. Union Switch & Signal appointed Errazuriz, Simpson and Company as sales agents in Chile and Bolivia from 1922 to 1928. When the agreement ended, they contracted with Beaver-Proud Engineering Company of Valparaiso, Chile, to take over as sole agents in Chile.

Between 1922 and 1926, Union Switch & Signal subscribed to 30 percent of the stock in Compagnia Italiana dei Segnali. The Italian company was absorbed by the Italian brake company, Compagnia Italiana Westinghouse dei Freni in

S. E. Gillespie

John Pressley Coleman

The Master's Apprentice

John Pressley Coleman, more often known as J. P. Coleman, was born September 8, 1865, in Allegheny City, Pennsylvania. He attended the public schools of Allegheny City, the John Way Academy in nearby Sewickley and entered Western University of Pittsburgh. In 1881, however, and just 16 years old, he had the unique opportunity to work for one of the country's geniuses, George Westinghouse, at Union Switch & Signal. Westinghouse was looking for a young man who could free-hand sketch the intricate mechanisms from the roughest of drawings that he so habitually scribbled on whatever paper was available. The drawings had to be accurate and to his specifications, so that the shop mechanics could start manufacturing the products quickly. Coleman left school and never looked back.

Shortly after the ambitious young Coleman started, he learned a valuable lesson about working for Westinghouse, which he relayed in a remembrance of Westinghouse after the latter's death in 1914.

He had been given a rough sketch of "a method for automatic block signal operation on a single track road," and his job was to "apply its disclosure to a series of successful block sections." Coleman had barely finished changing the diagram to avoid what he saw would have resulted in a dangerous situation, when Westinghouse entered the office in his usual rush.

"He attempted to trace the circuits," Coleman wrote. "I said nothing of the changes made from the original, hoping it would not be noticed or if noticed it would be accepted as something I was supposed to do…"Where is the sketch I made of this?" said he, turning on the stool and looking at me in obvious anger. Nonplussed,

I produced the sketch and stammered out why I had not followed it. His face flushed and his brown eyes grew darker as he glanced from sketch to drawing and from drawing the sketch, and then, apparently not satisfied with either, he rose and said, "Young man, hereafter when given a sketch by me, follow that sketch, right or wrong!" And with that he crumpled up the sketch and turned to go. He stood for a brief instant as if contemplating my youth and inexperience, and that with eyes still stern but a faint smile visible on his countenance, he looked at me and said in a kindly voice, "you know a good sailor never questions his captain's orders.'"

In 1888 he supervised the installation of the electrical part of the hydro-pneumatic system installed at the old Kansas City Terminal. In 1890, J. P. Coleman met signal engineer W. H. Higgins of the Central Railroad of New Jersey while installing the "first electro-pneumatic plant [system] of any considerable size" at the A, XA and RU towers in Jersey City. The following year, Union Switch & Signal installed the first electro-pneumatic interlocking that employed electrically actuated switch valves at the Jersey City terminal of the Pennsylvania Railroad.

He supervised the electro-pneumatic interlocking installation at both the Boston's North and South stations in late 1897, and three years later directed the 1901 installation on the Boston Elevated.

J. P. Coleman became Chief Engineer of Union Switch & Signal in 1911, and the same year he was designated Consulting Engineer. He was a prolific inventor and designer, and an authority in electro-pneumatic switches and signals. His patents included a 1914 railway signal that claimed, "a multipolar stator acts inductively on a rotor connected with the signal so as to move it through a limited angle on either side of a biased mean position."

On June 19, 1931, personal and business associates gathered at the Duquesne Club for dinner in Pittsburgh, Pennsylvania, to pay tribute to J. P. Coleman for his 50 years of continuous service with Union Switch & Signal. John P. Coleman died in 1949.

1928, to form the Compagnia Italiana Westinghouse Freni e Segnali. Union Switch & Signal exchanged their shares in this company for shares in the Westinghouse International Brake and Signal Company.

A. L. Humphrey handed the reins of Union Switch & Signal over to George A. (Augustus) Blackmore in 1932, when the former became Executive Director over both companies. Blackmore had spent his career at Union Switch & Signal, starting as an office boy in 1896 and rising up the ranks to Chief Clerk of the Engineering Department. He moved to the New York Sales Department in 1904 as a Chief Clerk and soon made Eastern Manager. In 1916, he returned to Swissvale as the General Sales Manager, and the following year he became Second Vice President. In 1922, he was made First Vice President and General Manager.

Swissvale employed about 2,500 people in 1923, not including field construction crews, which varied from 50 to 500 men, or their subsidiary Westinghouse Battery Company employees. Orders for 1923 approximated $10 million, of which $7.5 million represented signaling apparatus and $2.5 million the products of the forge department. In October of 1923, Blackmore announced that Union Switch & Signal had contracts to install Train Control on five railroads for a total of 111 miles of single track, 308 miles of double track and 277 engines; and another 15 miles of double track and six engines at two railroads for test purposes. The technology was catching on.

During the General Conference of the Westinghouse Air Brake Company and Associated Companies (including Union Switch & Signal), held October 15 to 17 of 1923, in Wilmerding, George Blackmore addressed

The Train Control laboratory at Swissvale moved into high gear to meet the new demands of five railroad contracts for Train Control installations in 1923. The contracts brought the railroads into compliance with a new ICC mandate.

The Westinghouse Castle, originally built as a combination executive offices and employee "club house," was rebuilt after a fire as the executive offices for Westinghouse Air Brake. After the 1917 merger of Union Switch & Signal and Westinghouse Air Brake, the building served both companies.

Photo credit: Joanne L. Harris

1923 General Conference of Westinghouse Air Brake Company and associated companies, including Union Switch & Signal.

By the mid 1920s, the era in which much of the railway equipment work could be accomplished in a blacksmith factory became history.

the subject of "The Signal and Train Control Situation." "The condition that regards train control is even worse [competition-wise]," he stated, "on account of the fact that spectacular deaths always attract the attention of inventors. There have been probably in the neighborhood of 5,000 patents issued in connection with train control…from the crudest mechanical devises to wireless control. We have never engaged in propaganda in favor of Train Control, but we have successfully tried and discarded various schemes (some of them on the same principle as those now being advocated by some of our competitors), until there was developed our present system."

Blackmore described one of the company's larger installations – the signaling system in the New York subways of the Interborough Rapid Transit Company, as consisting of 111 electro-pneumatic interlocking plants (systems) operating 2,312 signals and 975 switches. In addition, the system had 2,580 automatic stops. "The number of switch and signal operations exceeds 300 million a year, or nearly one million a day."

In 1923, Union Switch & Signal manufactured all four types of interlocking systems that were available at the time: mechanical, electro-mechanical, electric, and electro-pneumatic. Block signaling could be purely manual, controlled manual or automatic. Their installations had become so widespread that it was possible for a traveler to board a train in New York City and ride all the way to San Francisco under the protection of their automatic block signals every foot of the entire distance. One American railroad, with 7,000 Union Switch & Signal signals in service, reported that of 72 million signal operations in one year, they had only eight unsafe failures. That translated to nine million operations per one unsafe failure. Another prominent installation, the Pennsylvania Railroad's New York Terminal, comprised 11 complete interlocking plants

An operator in a central office managed his trains using centralized traffic control installation on this Central Railroad of New Jersey Between White Horse and North Branch.

Central Traffic Control is a system by which an operator in a central location controls train operation via signal indication.

having 516 working levers for the operation of 748 signals and 405 switches, and requiring over 2,800 relays, 512 track circuits, 1,700 electric lamps in signals, 15 miles of air pipe, and 1,723 miles of wire.

And yet, 72 percent of the passenger mileage in the Unites States had yet to be equipped with automatic signals. Nor was the potential in signal business growth limited to the Unites States. Orders from Japan, Australia, Canada and Chile indicated that railway leaders in those countries recognized the benefits of up-to-date signaling systems.

Engineering associated with railway signaling work had become highly specialized. Gone were the days when a master carpenter was in charge of signaling on the railroads and much of the work could be accomplished by a blacksmith. By the mid 1920s, the industry required signal engineers, electrical engineers, mechanical engineers and material engineers.

In 1923, George Blackmore was already referring to the future of Automatic Train Control as "a new era in train operation." As early as 1889, Union Switch & Signal acquired a patent for the first practical automatic stop which, he explained, "consisted of a glass tube attached to the train line, which would be broken by an arm extending from the signal should a train attempt to pass a "stop" signal." The first system of automatic stops successfully operated under rail traffic conditions in 1901 on the Boston Elevated and other mainline

George A. Blackmore

From Office Boy to President and Chairman of the Board

George A. Blackmore, was born in 1884 in Wilkinsburg, Pennsylvania, where he attended public school.

At the tender age of 12, he entered Union Switch and Signal's Orders Department in July 1896 as an office boy, earning $15 a month. The ambitious, young Blackmore completed a night school business course that earned him quick promotions from clerk to stenographer to Chief Clerk for the Engineering and Estimating Departments by the time he was 17 years old. He also completed several correspondence courses in engineering.

Blackmore was assigned to the New York City office in 1904 at age 20, when the first signal system was installed on the Interborough Rapid Transit subway. He was promoted to Assistant Eastern Sales Manager in New York in 1909, and became Eastern Sales Manager for the New York, Montreal and Atlanta offices two years later, responsible for sales and construction work. Blackmore briefly left the company to work as Vice President for Bryant Zinc Company, a manufacturer of gravity-type primary batteries for signal systems, from September 1914 to late 1915.

Blackmore returned to the Swissvale plant in 1915 as General Sales Manager. After the 1917 merger with Westinghouse Air Brake, he was elected Vice President in charge of Sales. He was promoted to Vice President and General Manager in 1922, elected to the Board in 1925. When Arthur Humphrey became Chairman of the Board in 1929, Blackmore succeeded him as President and General Manager of Union Switch & Signal. Blackmore, then 45, remained President of Union Switch & Signal until 1946. In the Fall of 1931, they sailed with Westinghouse Air Brake's Vice President C. A. Rowan on the S. S. Leviathan on a European tour to survey business affairs with associated foreign enterprises in Great Britain, Western Europe and Australia.

When Humphrey submitted his resignation to the Board of Directors in July of 1936, he was elected Chairman of the Executive Committee of Westinghouse Air Brake and Union Switch & Signal. George Blackmore succeeded him as President of Westinghouse Air Brake – the youngest man ever to fill this office other than George Westinghouse. Blackmore

was elected Chairman of the Board for both companies in 1940.

Blackmore was described as "quick to appraise the merits of any new engineering development or research achievement, and determine its commercial possibilities." Unlike Westinghouse, he took time out for recreation. George Blackmore, along with most Union Switch & Signal and Westinghouse Air Brake executives, was a founding member of The Millionaires Club at the Longue Vue Country Club, the private riding and golf club that A. L. Humphrey helped build in 1920. A brief article from 1943 stated, "the only things that make him forget business are boxing matches and fishing trips...or a detective story."

During the 1920's, Blackmore lived with his first wife Ethel in Squirrel Hill; by 1940 he lived in Perrysville with his second wife Mary and three children. The painful irony of war hit home with Blackmore in 1943. As he was leading Union Switch & Signal in the manufacture of airplane propellers and other aircraft materials, his 25-year-old son, First Lt. Harry R. Stengle, died in action as a fighter pilot in the European Theater.

Blackmore was also President and Director of Union Switch & Signal Construction Company, and from 1937 to 1940, he was President of Massey Concrete Products Corporation. He served on at least 17 organizations' boards in the United States, Canada and England, five of which were Westinghouse companies or affiliates, as well as the United States Chamber of Commerce and the Western Pennsylvania Hospital in Pittsburgh.

In accepting a bronze plaque commemorating his 50 years of service at Union Switch & Signal in July 1946, he stated: "Continuous devotion to a line of endeavor over so long a period has made the signal business so much a part of me that I hold no greater desire than to have a part in its continuing progress. While the company had been in operation for 14 years before I joined it, we were still in the pioneering stage. The company had about 400 employees. The machine work was done in an old roundhouse, and the office was a little building...I am more concerned with the possibilities of the future than with what has occurred in the past."

George Blackmore completed 52 years of service with Union Switch & Signal before his death on October 2, 1948, in Pittsburgh, only 64 years old. His plaque is on display at the Ansaldo STS USA corporate headquarters building.

H. A. Wallace

The CTC operator moved small levers on a control panel to send electronic messages to control the wayside signals and track switch mechanisms.

railroads. About 1912, the Union Switch & Signal installed another version of automatic stop or train control, called the "ramp" style, on the Lackawanna, the New Haven and the Pennsylvania trunk line railroads. These did not pan out, as they soon realized the ramps created clearance problems and derangements due to dragging equipment. Thus, in Blackmore's opinion, "The only satisfactory form of train control must be a system that abandoned 'intermittent' control and embodied the 'continuous' control of the train."

Before Centralized Traffic Control (CTC) was developed in the late 1920s, train operation was largely a group effort. The timetable established the schedule. The dispatcher hand wrote a train order and then telegraphed or telephoned each train's order to modify a schedule or add trains. Operators received the orders and relayed the messages to the train crew. The train crew operated switches manually. While everyone maintained diligence, there were too many potential points of failure. CTC technology sought to simplify the process.

Centralized Traffic Control is a system by which an operator in a central location controls train operation via signal indication. The operator controls the switches and signals over an extended territory, directs trains only by signal indication, operates wayside switches and crossovers electrically, and automatically receives visual and audible indication of the location of each train. The benefits of the system include a reduction in operating costs and improved train operation. When the dispatcher operates levers on the control panel, an electric message is transmitted via a communications circuit to a remote location in the CTC territory to control the wayside signals and track switch operating mechanisms. The vital circuits interlock switch and signal operation so that the system fails to a safe condition if circumstances are unsafe (for example, a train already occupies the block). The communication circuit is "non-vital" because a failure of the circuit will not create an unsafe condition. Using vital circuits only where they are essential provides the required safety measures while maintaining economical train operation.

The inventor of CTC was Sedgwick N. Wight, a General Railway Signal engineer with over 90 patents to his credit. His idea to consolidate switch and signal control at one location was originally called the "General Railway Signal Company's Dispatching System," but it was later generically renamed "Centralized Traffic

Control," or CTC. H. A. Wallace, a Union Switch & Signal engineer credited with over 100 patents, was a pioneer of early interlocking and CTC system design. His CTC patent, granted in 1922, fully integrated the system, giving the dispatcher control of wayside track switches and signals that were transmitted directly to the train crew. Wallace's co-worker, Clarence S. Snavely, saw the potential in modifying the Gill Selector, one of the products the company obtained through the Hall Signal Company acquisition, to work in the remote CTC system.

Union Switch & Signal installed its first CTC system in 1928 on the Pere Marquette Railroad between Mount Morris and Bridgeport, Michigan. The code line, with selector-controlled functions, cut down the number of wires required for the communication circuit, which at this time was a real benefit with the copper shortage at that time. The line extended 19.8 miles and was more advanced, but less flexible, in regard to the switch positions indications. Another milestone was reached in 1929 when Union Switch & Signal installed an all-relay, circuit-code system of CTC at Eagle Bridge, New York, on the Boston & Maine Railroad.

One of the earliest known photographic references to Union Switch & Signal as "the Switch" was in 1914. An early written reference was in the March 1928 issue of the WABCO News.

In 1924, General Railway Signal and Federal Signal companies merged, granting General Railway Signal a 44 percent royalty base according to the 1916 Interlicensing Agreement. In September 1925, Union Switch & Signal acquired the Hall Signal Company's assets except for their corporate franchises, gaining them a royalty basis of 56 percent of the Interlicensing Agreement. Among the assets was the Hall Signal Company's pioneering searchlight signal. This new signal provided day/night aspects that featured a single light source, instead of the previous three-lamp systems of color light signals or seven in position light signals. The low-wattage lamp provided a beam that used colored, movable roundels to define the signal aspect. Union Switch & Signal maintained the operations of the former Hall Signal Company Garwood, New Jersey, plant for less than one year, at which point they disbanded the workforce and sold the real estate. From 1925 to 1965, Union Switch & Signal and General Railway Signal controlled 90 to 95 percent of the United States market for railway signaling equipment.

In 1926, George Blackmore informed his engineers that they had two weeks to develop and ship a sample of coded cab signaling equipment to the Pennsylvania Railroad for an order he had already sold. Herbert "Skip" Wallace had initiated the idea of using a code in the track for cab signal purposes, but nothing had yet been produced. Four men were assigned the four parts of the puzzle. Clarence S. Snavely was assigned the decoding equipment; George W. Baughman the filter equipment; Herman G. Blosser the amplifier; and L. V. Lewis the box and arrangement of the unit. These men were all inventors in the company. Blosser collected 40 patents to his credit by the time he retired from Union Switch & Signal in 1959; George Baughman had about 111.

Frank "Nick" H. Nicholson, a Union Switch & Signal train control engineer, later explained in a speech that, "The code system is so called because the signal indications aboard the locomotive are controlled by coding the track circuit current; that is, by alternately opening and closing the circuit supplying the electric current to the track rails. The number of interruptions per minute, or the "code," determines which of the four signal indications will be displayed in the cab of the approaching locomotive." In 1930, the Pennsylvania Railroad put in Union Switch & Signal's first coded continuous system that showed danger in the cab by visual indication and whistle-blowing only; no automatic brake-related equipment was used.

Union Switch & Signal and General Railway Signal entered an interlicensing agreement in 1926 to cover their respective Canadian patents. The next year, Union Switch & Signal introduced the copper oxide rectifier, and gave the Westinghouse Electric company exclusive license in the United States and non-exclusive rights in any country where Union Switch & Signal had the right to grant licenses. The Westinghouse Brake and Saxby Signal Company also received a license under the same patents to apply to the French, Italian, Indian and Australian companies.

The ICC wrote in its November 27, 1928 report: "Cab signals are without a doubt an important development in the art of signaling. They place the signal indication immediately in front of the engineman where it cannot be obscured by snow, fog, smoke, or other obstructions, and, where a combination of visible and audible indication is used, it is without a doubt a valuable addition to the signal system." The ICC also recognized that signaling apparatus standardization was not possible because of the varying requirements of usage, application and control. As Blackmore once stated, "…in the case of one electrical instrument we manufacture, there may be a call for any one of 221 possible assemblies." In the same train control report, the ICC announced that there would be no more orders calling for installations of train control at this time. Apparently, the commission considered that its previous two orders accomplished sufficient development of automatic train control. This put the decision making back in the hands of the railroads as to whether or not they wanted to add more train control, cab signals or other safety measures on their lines.

In 1927, the Westinghouse International Brake and Signal Company was organized and took over the foreign holdings of Union Switch & Signal and Westinghouse Air Brake. Union Switch & Signal exchanged its shares in the Westinghouse Brake and Saxby Signal Company and Compagnia Italiana dei Segnali for 1.3 percent of the shares in the newly formed company. They transferred their shares to the Swissvale Corporation, the wholly owned subsidiary of Union Switch & Signal that held their investments.

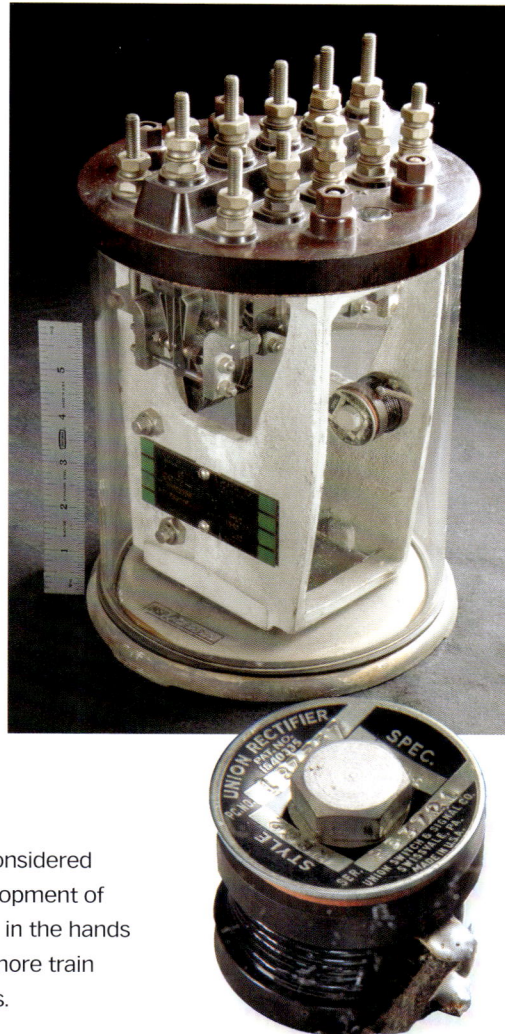

The copper oxide rectifier, invented by Lars O. Grondahl and P. H. Geiger in 1927.

In 1928, Union Switch & Signal signed a 20-year patent and manufacturing license agreement with Kabushiki-Kaisha Kyosan Seisakujo (Kyosan Engineering Works, Ltd.), granting the Japanese firm rights to manufacture Union Switch & Signal products in the Japanese Empire, Formosa, Korea and the Japanese-controlled Railway Zone in Manchuria. Union Switch & Signal received 20 percent of the company's stock and other dividends. The agreement was the culmination of a long controversy with Kyosan, which, after manufacturing copies of the Union Switch & Signal designs in 1921, secured the majority of the Japanese market.

In 1927, Herbert L. Bone came to work for Union Switch & Signal. He was the company's most prolific inventor, who amassed a collection of 149 patents during his tenure between 1927 and 1956.

Nick Nicholson presented a paper, "What is the Future of Train Control?" at the 1929 Air Brake Association convention in Chicago, in which he proclaimed, "The ideal train control system will have two functions: first, to provide information by which the engineman can handle his train properly; and second, to stop the train automatically or reduce its speed in the event that the engineman is unable, or for any reason, fails to handle his train properly."

More than 9,000 (over 14 percent) locomotives and electric cars in America were now equipped with Automatic Train Control, operating on over 20,000 miles (nearly 7 percent of the country's track mileage) of equipped track in 35 states. There were "four distinct types of automatic train control equipment" in use, either in compliance with orders of the ICC or voluntarily: Automatic stop of the continuous type with forestaller; Automatic stop of the intermittent type with forestaller; Speed control of the continuous type; and Continuously-controlled cab signals.

"The second type does not prevent a reckless or a careless engineman from taking chances after passing the restrictive signal," stated Nicholson. "This hazard is eliminated when a continuously-controlled cab signal is associated with the train stop system, because no engineman, in his right mind, would drive on toward the next signal at stop, while the signal in his cab told him positively and continuously that the next signal had not cleared. Maximum protection is obtained...when traffic conditions in advance become more restrictive after the train has passed a signal location. With the continuous system, under such conditions, the cab signal changes immediately and the brakes apply automatically if the prescribed action is not taken by the engineman."

An analysis made of public reports on 15 outstanding collisions in the United States between 1923 and 1927 indicated beyond all reasonable doubt that 14 of them would have been avoided had the engineman and fireman received the information provided by the continuous system of cab signaling. And the railroads responded. About one in every seven locomotives in this country became equipped for automatic train control or cab signaling.

The commercial race for train control was on, in spite of the dreary fact that the Great Depression shriveled up investment funds. Salaries and hours were cut in half at Union Switch & Signal, but the plant never closed, and the research and development continued. George A. Blackmore, by then President of the company in 1929, along with M. L. "Doc" Gray – Vice President of exporting, and W. H. Cadwallader, General Manager, encouraged engineers like Wallace and Snavely to advance the CTC program.

The American Railway
Association added "Think"
signs to Union Switch &
Signal's HC-2 grade crossing
poles in 1925 in an attempt
to reduce accidents.

A Union Switch & Signal H2 Searchlight signal pierces the darkness along the BNSF Railroad in Robertsville, Missouri.

Photo credit: © Zachary Gillihan Collection

8   Expansion Amidst the Great Depression

Many companies floundered and failed in the mire of the Great Depression, yet some, like Union Switch & Signal, managed to keep their heads above water throughout the tough times. In 1929, George Blackmore had taken the reins as President of Union Switch & Signal, and throughout the next decade, he pushed engineers like Crawford E. Staples, also known as "Tacky," Nick Nicholson and Charles "Chick" B. Shields to push forward with their work on the coded track circuits. Staples and Nicholson had completed the apprentice program at the company, but all three were destined to do great things. Staples came from the train control section, where he had worked on the coded wayside system under his supervisor, Herman Blosser. Later known as the "world's leading expert in coded track circuits," Staples built a resume of 64 patents during his career at Union Switch & Signal.

Frank "Nick" H. Nicholson

Nick Nicholson had started with the company in 1911, but left for several years to work as a circuit designer for the Interborough Rapid Transit Company in New York, and later as a signal inspector on the New York, New Haven & Hartford railroad. He returned to Union Switch & Signal to work with Dr. Grondahl on projects that led to the copper oxide rectifier discovery in 1926. Nicholson later took on the position of Chief Engineer.

Chick Shields discovered an early problem with coded track circuits when they performed an install in Lewiston, Pennsylvania. The relay was supposed to go down when they put the battery on one end of the track circuit and took it back off, but it did not. They soon realized that the ballast under the rails worked like a storage battery. When the current was put into the tracks, the ballast picked it up and stored it, and the current could not be turned off. This discovery by Shields paved the way for future coded track circuit technology. The engineers also learned that freezing

Members of the American Railway Association Safety Group of the ICC, including Arthur Humphrey and George Blackmore, at Wilmerding, November 29, 1932.

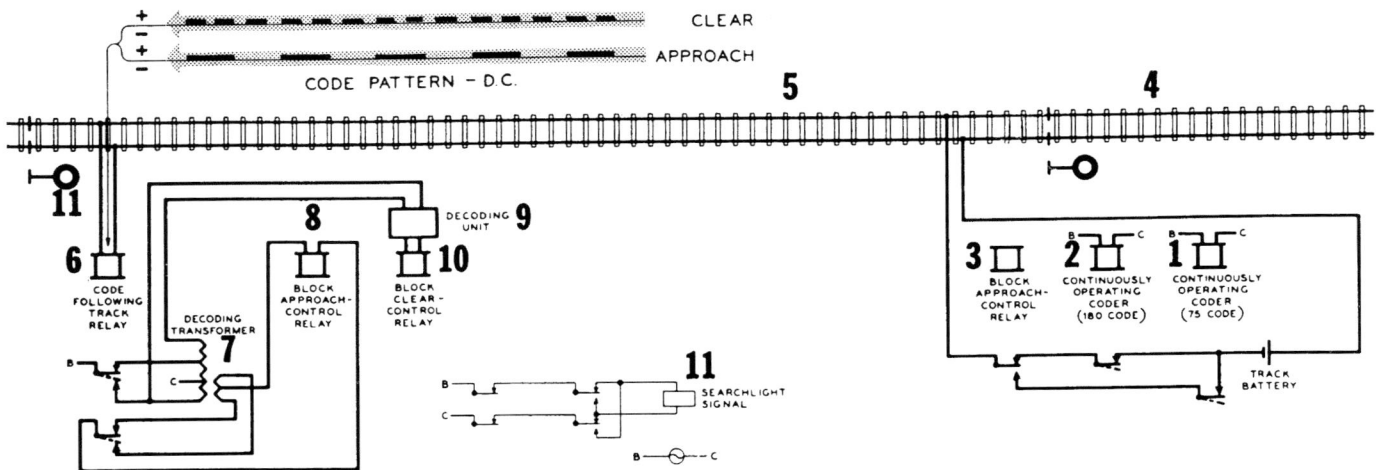

CODE PATTERN – D.C.

Typical three-indication (red, yellow, green) coded track circuit. In 1935, the company developed the first coded track circuit signaling, which they named the "Union" Coded Track Circuit Control.

Charles "Chick" B. Shields

temperatures often caused coded track circuits to malfunction. When the condensers became too cold, the loss of energy changed the capacitance, causing them to fail. The engineers created a workaround by installing a light bulb in each signal case to keep the condensers warm enough to function properly.

In 1933, the first three- and four-indication wayside and cab signaling was installed in electrified territory. The following year, Union Switch & Signal installed its first three- and four-indication wayside and cab signaling in steam territory, and its first multiple track territory with reverse running on one track. In 1935, the company developed the first coded track circuit signaling, which they named the "Union" Coded Track Circuit Control. The system provided for a longer track circuit, and passed information through the rails, eliminating the need for line wire circuits. The extensive savings on copper and other materials during World War I shortages led to wider use of coded track systems. By May of 1936, they had 500 miles of coded track circuit in service, all on the Pennsylvania Railroad.

Toward the end of the decade, the first 11,000-foot track circuits began using primary or storage batteries. The company applied their first coded reverse circuit for approach lighting of wayside signals and approach application of cab signal energy. They finished the decade in high gear as the coded track circuit continued its pace with three more "firsts" in 1939: three-indication wayside signaling for steam territory employing 11,000- foot track circuits; the use of a tuned alternator as a standby for supply of cab signal energy; and coded track circuits used as "detector" track circuits in interlockings.

Interlocking machines grew to all-time sizes in the 1930s. Union Switch & Signal installed a 115-lever Model 14 machine in the Harris Tower for the Pennsylvania Railroad in 1930. This machine controlled the west end of the Harrisburg station complex. A 127-lever unit at State Tower was installed around 1937 to control the east end. Union Switch & Signal produced only about 200 of the Model 14 machines between 1916 and 1959, but over 10 percent of the machines boasted more than 100 levers! The largest Model 14 was a 367-lever unit for Pittsburgh's PITT tower. The last Model 14 manufactured was a small, 37-lever unit for the Pennsylvania Railroad's Queens Village station on the Long Island Railroad in 1959.

Streamlined trains were designed in the 1930s to seduce passengers back on the rails after the Great Depression had diminished ridership. Burlington's famous Zephyr train was one of those new trains. The sleek Zephyr set a world record in May 1934 when it sped from Denver to Chicago – 1,015 miles – in 13 hours and 5 minutes –averaging 77.6 miles per hour, and peaking at 90 miles per hour at Lincoln, Nebraska. It arrived at Halsted Street in Chicago at 7:09:40 p.m. Central Time, but it did not stop! It barreled on through to make its dramatic entrance at the "Wings of a Century" stage at the Century of Progress Expo, Chicago's World's Fair that ran from May 1933 to 1934. In spite of the glamour and fanfare of these new streamliners, though, they remained confined by the current mode of dispatching.

Three years later, in 1937, the Burlington Railroad installed the first complete operating subdivision of Union Switch & Signal Centralized Traffic Control (CTC) equipment. A 112-mile subdivision handled Denver traffic from three Burlington routes. The Chicago-Omaha-Lincoln-Denver main line and the St. Louis-Kansas City-Denver line joined at Oxford Junction, Nebraska, and ran westward on single track to Denver. The Sterling division, north to Alliance, Nebraska and Billings, Montana, intersected from the North at Brush, Colorado, forming a major 88-mile bottleneck into Denver. Rather than double track, the Burlington installed the "Union" system of Centralized Traffic Control. The new system controlled 20 sidings as well as the junction at Brush.

A 1939 Union Switch & Signal Bulletin described the "Union" Centralized Traffic Control: "[It] permits handling of freight trains as soon as they are ready, without any lost time in the transmission of train orders, as the mechanics of transmittal are reduced to the simple movement of a small lever without requiring the intermediary action of another person." The dispatcher observed train movements on his illuminated track model, which displayed track circuit occupancy and switch position. The track model could be floor mounted or be suspended from the ceiling, as space dictated. He judged the trains' respective speed and issued orders accordingly by signal indication at the points of action. A permanent record of the treks that each train made was printed out on an Automatic Train Graph.

Union Switch & Signal first tested the use of transistors on a railroad with an automatic track gang warning device developed in conjunction

Union Switch & Signal installed a 115-lever Model 14 machine in the Harris Tower for the Pennsylvania Railroad in 1930.

Photo credit: © Rich White II

The "Union" Centralized Traffic Control bulletin described a 1937 installation on a 112-mile subdivision of the Burlington Railroad. "All trains from Akron to Denver operate entirely by signal indication."

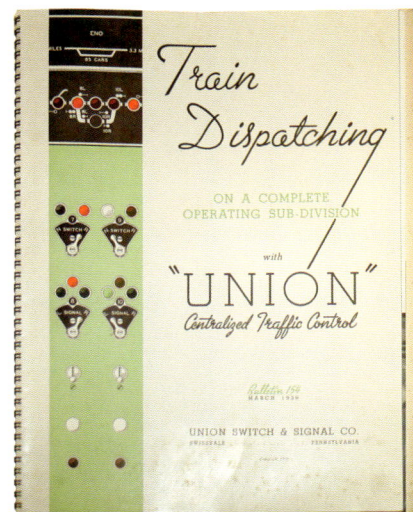

Operator at a CTC control board at the Central Railroad of New Jersey in 1938.

with Wynn Salmonson of the Pennsylvania Railroad. This device consisted of two boxes. One was placed upstream of the point where the protection was required and connected to a power source and to the rails. The second box was placed across the rails where the workmen were stationed and was equipped with a bell and flashing lights. The first box created an audio frequency in the vicinity of 1,000 Hz and applied it to the rails. The second box detected the audio frequency and energized a relay. When a train occupied the track between the boxes, the relay was de-energized and the bell rang and the lights flashed, proving that audio frequency could be overlaid on existing track circuits without interference.

The Style SLV-13 vane relay, circa 1940, was used to detect track occupancy.

Not only did Union Switch & Signal make great strides in the coded track circuit and transistor arenas, but they were equally prolific developing and manufacturing a vast array of new and improved products in the 1930s. There was a series of vane relays for automatic train control; polarized and neutral relays, as well as specialized relays for approach lighting applications (the DNL line); thermal relays with an adjustable time delay to release the approach locking of a switch machine; and an ANL-40 Transformer Relay that was used for on-off detection. A Style "DT-10" DC Time Element Relay featured an oscillating armature used to create rotary motion via a ratchet wheel and panel, resulting in contact closure after time delay; this relay was used for time locking of switch machines.

Union Switch & Signal's pedestal-type dwarf signals – so called because of their relatively low height – became popular at terminals and other interlockings with short distances between signals. The pedestal type was 7 feet high and only 16 inches wide, and was ideal for locations where congested track layouts limited clearance or as an alternative to the higher cost overhead signal support structures.

The Style M-22 Dual Control Switch and Lock Movement permitted the selector relay to be thrown to either extreme position

Union Switch & Signal introduced a number of crossing signal styles in the 1930s.

regardless of the relative position of the motor mechanism, with the exception that the hand throw lever had to be in the extreme normal or reverse position. Other safety products built in the mid- to late-1930s included portable rectifiers, the hand-operated T-20 and T-21 switch mechanisms, and the PN-50 plug-in relay that allowed the railroads to safely replace a relay unit without disturbing the external wires. Many of these product lines created the foundation of the company's core products today.

As the popularity of the automobile grew, so did the number of grade crossing fatalities. The first automatic electric motor-driven crossing gate was introduced in 1936 on the Illinois Central by adapting a semaphore signal mechanism to drive a wooden gate arm. In 1937, when only three passengers were killed in train accidents, 1,875 were killed at grade crossings. Accidents at highway crossings became one of railroading's most frequent and tragic occurrences, as drivers naively sought to "beat the train." Union Switch & Signal introduced a number of crossing signals, including the "HC-7" Highway Crossing Signal with "Backlight," which enabled motorists and pedestrians to see the signal that otherwise might be obscured from view.

Internally, Arthur Humphrey was made Executive Director of Westinghouse Air Brake and Union Switch & Signal in 1932. Charles A. Rowan, George Blackmore and S. G. Down were elected as directors of Union Switch & Signal, creating identical boards of directors for the two companies. A notable increase in plant activity at Westinghouse Air Brake increased Humphrey's workload to the point that he requested to submit his resignation as Chairman of the Board at the Board of Directors meeting on July 17, 1936. Humphrey was elected Chairman of the Executive Committee of both Westinghouse Air Brake and Union Switch & Signal. Charles A. Rowan was elected Chairman of the Board of both companies, assuming general supervision of the Companies' fiscal affairs. George Blackmore was elected President of the Westinghouse Air Brake, while retaining his position as President of Union Switch & Signal.

Arthur Luther Humphrey

A Man with a Resolve of Steel

Arthur Luther Humphrey was born in Buffalo, New York, in 1860, the youngest in a family of eight children. When he was less than a year old, the family moved to the midwest. He worked as a drug clerk when he was 14 in Plattsmouth, Nebraska. By the time he was 17, he was hired as a machinist's apprentice on the railroads. It was in 1877 when A. L. Humphrey was first introduced to the Westinghouse Air Brake Company. He assisted in the installation of the first air pump that was applied to a locomotive at Plattsmouth, Nebraska.

He advanced through the ranks and accepted a position in 1888 with the Colorado Midland Railroad. While living in Colorado, he became a member of the Colorado House of Representatives from 1893 – 1895, and was Speaker of the House in 1895. He gave many addresses at colleges and universities although he was largely self-educated. Humphrey returned to railway service, however, on the Colorado & Southern in 1899, and then went to the Chicago & Alton in 1903, as superintendent of motive power.

If a man's character is revealed by his choices, this paraphrased except from Arthur Humphrey's autobiography offers an indication of his tenacity and drive:

A terrible blizzard developed over the Continental Divide on January 1, 1898. Two days later, Humphrey led an expedition out of Leadville, Colorado, consisting of their only rotary snow plot, three locomotives and a caboose, "to extricate three passenger trains and several freight trains that were stalled in the snow over the mountain range." When they cleared the tracks and relieved every train, the crew "proceeded westward to open up the road on the west side of the mountain." An avalanche made their retreat impossible. Humphrey had 100 laborers in the caboose, and he set them to work chopping wood and shoveling snow into the tenders to keep the train moving once they cleared the track. He took over the plow operations, but the weather worsened and he soon had 100 hungry, agitated men with no food, coal or water.

The conductor became mentally unbalanced when he realized their predicament, and had to be restrained. Humphrey took charge of the laborers when the foreman also became unnerved by the situation such that he, too, became helpless. The laborers "rebelled and mutinied," refusing to obey Humphrey, stubbornly retreating to the warmth of the caboose. Humphrey knew their lives were at stake. He picked up a track pick and broke the pick off, threatening the men with the remaining club if they refused to work. The life-saving ruse worked and they all left the caboose. Humphrey locked the door and would not let them back in until the next day, when the plow was finally restored to the track. With a full supply of wood and water, they attempted to move, but the icy tracks continued to send the plow off the rails. By then, the men realized the seriousness of their predicament, and they proceeded to chop wood and shovel snow without coercion. They finally reached a section house where they found flour, beans and coffee. When the weather abated, they laid in full supplies for the return trip, but were once again cut off by slides and drifts. Wearing snow shoes, Humphrey

92

walked back to the section house for help, and sent the men back as well until replacement parts arrived. The crew arrived back in Leadville after 24 days absence.

Humphrey was still working with the trains when he received a letter from Henry Westinghouse, Chairman of the Westinghouse Air Brake Company, regarding an "improvement" that a cabman had discovered that was actually detrimental to the safety of the brake system. Gradually, Humphrey began to meet the people at the Air Brake Company, including E. M. Herr, then the General Manager for the Westinghouse Air Brake Company. Herr offered him a job as District Manager in Chicago in 1903. Humphrey began to visit the Wilmerding offices and soon met George and Henry Westinghouse. When Herr left the company, Humphrey moved into the position of General Manager in 1905, and rose to Vice President in 1910.

In 1917, when the United States entered the war, Humphrey converted the Union Switch & Signal Company shops into an airplane engine factory. Not surprisingly, he was a member of the Labor Advisory Committee for the Council of National Defense, and wrote a paper on "Railroad and Industrial Efficiency in War – How to Make the Most Effective Use of Forces and Equipment in the Emergency That Confronts the Nation." After the war, Humphrey became interested in commercial aeronautics, co-founding the Pittsburgh Aviation Industries Corporation and the Pennsylvania-Central Airlines.

In 1917, the Union Switch & Signal Company merged with Westinghouse Air Brake Company, and Humphrey, first vice-president and general manager of the Westinghouse Air Brake Company, was elected president of the Union Switch & Signal Company, assuming executive responsibility of both offices. He also served as President and director of the Westinghouse Traction Brake Company, Chairman of the Board of the Westinghouse Pacific Coast Air Brake Company and the Westinghouse Union Battery Company.

In 1942, Westinghouse Air Brake (WABCO) presented the Pittsburgh Chamber of Commerce a portrait of Arthur L. Humphrey. George A. Blackmore, president of Westinghouse Air Brake and former President of Union Switch & Signal, said in his address: "An amiable and charming personality… was apparent in his business dealings. But above all, he possessed an inherent ability as a real executive – a genius in the art of getting things done. I might recall that during the period in which the Union Switch & Signal Company was engaged in extensive manufacture of war material, the plant was demolished by fire. Mr. Humphrey set to work immediately to erect a modern fire-proof plant of more than twice the original size and completed all contracts on time."

He served two terms as President of the Chamber of Commerce in Pittsburgh from 1923 to 1925. During that time he launched an extensive campaign that increased membership from 3000 to over 6000, and transformed the group from local to national status. His last article, "The Battle Against Ignorance," appeared in 1932 in "Nation's Business."

When Arthur L. Humphrey died at home in Edgewood at 79, he was an executive in 15 companies, many of them divisions of Westinghouse; five banks; Pennsylvania-Central Airlines and an industrial aviation company, and was a trustee of the University of Pittsburgh.

William H. (Henry) Cadwallader, General Manager since 1914, was made Vice President in April of 1936, and held the positions of Vice President and General Manager until he retired at age 69 on December 31, 1945. He was elected as a Union Switch & Signal director from 1940 until 1948. Like Blackmore, he had humble beginnings as a career employee at the company, entering the firm in 1891 as a blueprint boy, moving into a clerk position three years later. Although he was a signal engineer, he never took out any patents.

The Franklin Institute awarded the Elliott Cresson Medal to Union Switch & Signal for remarkable advances in train control technology in 1934. The medal was awarded for "discovery of original research adding to the sum of human knowledge, irrespective of commercial value; leading and practical utilization of discovery; and invention, methods of production embodying substantial elements of leadership in their respective classes, or unusual skill or perfection of workmanship."

William Henry Cadwallader

In May 1932, the original building where Westinghouse first opened Union Switch & Signal at the triangle site on Duquesne Way and Garrison Alley was razed. Within the walls of the 50-some odd years of service, it had seen the birth of many an invention, from railroad safety equipment to the electrical inventions of Westinghouse Electric, the Nernst Lamp Company and finally, the R. D. Nutall Company, a Westinghouse subsidiary. The building, with all of its history, was sacrificed for yet another parking lot.

On the company's international scene, the 5-party agreement expired in 1935, and in 1936, the International Company was dissolved, distributing its assets pro rata among its three shareholders. The Union Switch & Signal holdings were again transferred to the Swissvale Corporation, which sold the shares to the Westinghouse Air Brake. In 1937, Andersen Meyer & Company was appointed sole sales agent in China, giving them permission to manufacture under Union Switch & Signal's licensed designs. The same year,

The Franklin Institute awarded the Elliott Cresson Medal to Union Switch & Signal in1934 for their advancements in train control technology.

an inter-company territorial agreement was entered into by the German, Italian, French, Belgian and English companies, defining their territories for the four fields of brakes, signals, rectifiers and heating devices to provide cooperation among the companies. In 1939, Fonseca-Almeida was appointed sole sales agent for Union Switch & Signal in Brazil to deal with the Government Railways for a five-year term.

In 1934, the United States Government financed a program to electrify 109 miles of line between Wilmington, Delaware and Washington, D.C. The Pennsylvania Railroad, which owned the line, re-arranged the automatic block signal system, using position-light signals spaced for maximum train speeds of 90 miles per hour. The new coded track circuit system controlled the wayside signals and the continuously controlled cab signals without using line wires. The line was placed in service in the fall of 1934. The line consisted of 27 interlockings along sections of two-, three- and four-track railroad, including 314 track miles of signaled main line. Eighteen of these track miles were signaled for dual-direction train operation. Some turnouts and crossovers were changed to accommodate higher train speeds, requiring changes in the interlockings and re-spacing of the home and distant signals.

The first Union Switch & Signal building at Garrison Alley was razed in May 1932.

The new coded track circuit system on the Pennsylvania Railroad (PRR) controlled wayside and cab signals without using line wires. Here, the PRR electric P5 4743 traveled from New York to Washington on April 28, 1933.

Union Switch & Signal furnished car retarders for 19 classification yards, including this Allentown Yard on the Central Railroad of New Jersey, and six car dumper installations within the first 14 years after the car retarder was made practicable in 1924.

Three-aspect, two-block signaling replaced the semaphores on automatic blocks that averaged 8,000 feet in length.

Within 14 years after the car retarder was made practicable in 1924, Union Switch & Signal furnished car retarders and associated apparatus for 19 classification yards and six additional car dumper installations. The car retarder took the place of the hand brake, saving many injuries to yard workers. By 1932, Union Switch & Signal had installed freight classification yards systems in numerous locations, including the Pitcairn Yard on the Pennsylvania, the Russell Yard on the Chesapeake & Ohio, the Allentown Yard on the Jersey Central and the Sharonville Yard on the Big Four, the Marion Yard on the Erie and the Potomac Yard on the Richmond, Fredericksburg & Potomac Railroad.

In 1936, the "Union" Inert Car Retarder came forth as an inexpensive retarder especially suited for classification yards that handled all the same type of freight – either all empty or all loaded freight cars. It applied a constant amount of retardation to all cars for given weather or traffic conditions.

Classification yards such as the Hampton classification yard in Scranton, Pennsylvania, required 40 classifications for its average of 600 cars daily when the nearby anthracite mines it supported were in full swing. Union Switch & Signal provided electro-pneumatic power switches, signals and retarders for this

gravity-operated, or "hump" yard. The hump yard at Clearing, Illinois, with the Model 31 electro-pneumatic retarders, processed as many as 6,000 cars a day. The Enola hump yard in Harrisburg, Pennsylvania, could classify about 1,200 cars in eight hours, working 45 minutes per hour. One 110-car train was classified in a mere 23 minutes. Enola yard was the "neck of the bottle" for all eastbound freight traffic on the entire Pennsylvania system. The new facilities were placed in service in December 1937, and in the first 15 days of May in 1938, 29,497 cars were "humped!"

Operators controlled electro-pneumatic switches, signals and retarders from the Pitcairn classification hump yard tower at the PRR to classify train cars faster and more efficiently.

Although Union Switch & Signal and Westinghouse Air Brake made great inroads in the CTC and car retarder business in the 1930s, American companies were suffering in the Great Depression, and the railway supply industry was not spared. Union Switch & Signal operated at a loss in 1938 due to the "sharp reduction in signaling installations by the railroads," with 59 percent fewer sales than in 1937. Even Westinghouse Air Brake sales for 1938 dropped 57 percent. In reality, this was to be expected. After all, the ICC mandate for cab signaling had ended in 1929, and the entire country was battling one of the difficult economic decades in its history.

Employees at Union Switch & Signal contributed ten percent of their wages to support the World War II war bond drives.

9   From Rails to Bomb Tails

Just as the United States was beginning to emerge from the Great Depression of the 1930s, it was faced with a second great war, World War II. War materiel production was underway at both companies by the time George Blackmore became Chairman of the Board for both Westinghouse Air Brake and Union Switch & Signal after Charles A. Rowan died on September 13, 1940. Under the leadership of Blackmore and William H. Cadwallader, who was Vice President and General Manager of Union Switch & Signal from 1940 to 1946, the company set out to provide the necessary support to the United States and its Allies.

Union Switch & Signal was a prime manufacturing candidate during World War II, as it had the available manufacturing space, machinery and trained personnel. More specifically, it had met World War I production schedules. The Company maintained "round-the-clock" schedules, making full use of its assets to support the war effort from early 1940 through mid-1945. In all, the company produced more than 43.5 million war materiel items, including bomb casings, shells, parts for airplane propellers, devices for radio controlled aerial bombs and Colt .45 automatic pistols. Additionally, they forged more than 250,000 additional units and piece parts.

The challenge to quadruple pre-war output within the existing plant required an increase in personnel from the pre-war figures of 1,800 employees to a war-time peak of more than 6,350 workers across three shifts, seven days a week. The Company reallocated space throughout their facilities to the best advantage, including a guarded Special Development Department for "Secret" or highly secured manufacturing processes. Their success was evident when, during a 1942 visit to the Swissvale plant, Lieutenant General William S. Knudsen stated, "What I like about you fellows is that you took your old plant and made a *real* manufacturing institution out of it."

Orders rolled in from the Pittsburgh Ordnance District, the United States Navy, the Army Air Forces, the United States Treasury Department and the British Purchasing Commission, along with manufacturers' subcontracts. Beginning in late 1940, Union Switch & Signal began machining bodies for 100,000, 40-pound, high explosive aircraft bombs for the British Purchasing Commission. At 10,000 bomb casings a month (half of the contract was sub-contracted to Westinghouse Air Brake), the entire contract was completed in six months, exceeding delivery schedules. From there, they produced tail locating studs that anchored lock nuts on the bomb fins of 100- and 200-pound bombs;

The Azon-2 high-angle, dirigible bomb with flare was one of the 43.5 million items manufactured at the Swissvale plant from early 1940 to mid-1945.

Interior of Union Switch &
Signal ammunition manufacturing building in Swissvale.

The standard hub spiders
and housing barrels were
two of the most intricately
machined pieces of war
materiel produced in Swissvale for the war.

component parts for chemical mortar shells; and 20mm Oerlikon shells
for the British and United States Navies. Union Switch & Signal manufactured over 35 *million* shells in just 44 months. Contract after contract
poured in, and the Company was soon building breech block parts for
U. S. Army Howitzers and parts for M-3 medium tanks. An order for
302,725 high explosive shells was produced and shipped in less than
three months. Though the order was cancelled, the Ordnance District
representatives said that the Company "was producing the best shell
of the type in this district."

Two of the most intricately machined pieces of war materiel
produced were the hub spiders and housing barrels for the 23-E-50 and
33-E-60 hydromatic airplane propellers. These propellers were utilized for
a variety of military aircraft, from advanced training planes to the latest
Superfortress. Union Switch & Signal manufactured more than 250,000
parts working around the clock, seven days a week, to satisfy the United
Aircraft Corporation's (now United Technologies Corporation) contract.

Potentiometers and gear drawer assemblies for the Signal Corps
Radar Continuous Sight (also known as the SCR-547 plane-locating instrument) were manufactured and assembled in the Special Development
Department between mid-1942 and early 1943. Western Electric Company
and Bell Telephone Company Research Laboratories jointly developed

The classic Colt .45 automatic pistol became a highly prized collector item after the war. Union Switch & Signal manufactured some of the parts and assembled 55,000 of the model 1911-A-1 pistols for the United States Army in 1943.

Photo credit: Joanne L. Harris

All-women crews produced an average of 1,200 dial wheels per day for Teletype machines.

this range-finding device, popularly known as the "Mickey Mouse" due to its shape. The SCR-547 was capable of locating aircraft within a 20-mile range, projecting it image onto a screen, and synchronizing gunfire upon the aircraft.

One of the four prime producers in the United States, Union Switch & Signal produced 55,000 Colt .45 caliber automatic pistols, designated by the Army as M-1911-A-1, between February 15 and November 27, 1943. Union Switch & Signal sent a group of 12 men to study production methods and techniques at the Colt Patent Fire Arms Manufacturing Company in Hartford, Connecticut, so they could head up manufacturing of their 16 assigned components and weapon assembly. Of the pistol's 54 component parts, 16 were manufactured in the plant. The remaining parts were delivered to Union Switch & Signal as "partially built" by outside contractors, or they were purchased from outside sources. Production took up nearly two floors in one building; additional space in another building was used for heavy machining operations.

Union Switch & Signal trained women to perform complicated assembly and wiring tasks during the war. All-women crews produced an average of 1,200 dial wheels for Teletype machines daily to satisfy an order of over 594,000 units in 28 months; they assembled 17,210 sets of

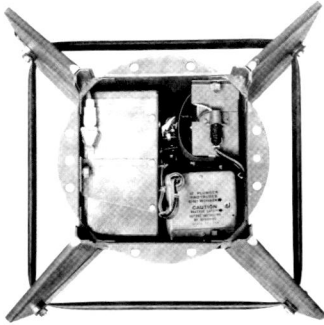

Workers crafted the tail assembly for the Azon-1 (Azimuth Only), a high-angle dirigible bomb with radio-control capabilities. (rear, or flare, view of device)

an ignition shield harness for airplanes equipped with Pratt & Whitney motors to eliminate motor ignition noises from the plane's radios. The women built one version for the Army Air Forces, and another for the Ford Motor Company.

The United States Army Air Force wanted the "Azon"(AZ-1) high-angle, dirigible bombs in World War II to increase accuracy of heavy bombs (1,000 pounds or more). Union Switch & Signal built the tail assembly for the bombs with radio-control capabilities and refined the design for higher efficiency. The Azon increased precision bombing 20 to 40 times over standard bombs for narrow bridge bombings. The name "Azon" (from "azimuth" and "only") indicated that an Azon-equipped bomb could be "controlled to the left or right along the azimuth in a plane parallel to the flight of the bombing plane during free fall." The redesign, assembly and test work was done under 24-hour guard in the Special Development Department, with restricted admittance. Union Switch & Signal also developed and manufactured the "Razon" (VB-3) in August of 1945. The Company engineered, tooled and produced 150 units of the Razon – a tail attachment for an airplane bomb of 1,000-pounds – to make it "dirigible;" that is, capable of being controlled while in flight as well as in azimuth and range.

The World War II work accounted for a large share of Union Switch & Signal business in 1942. At the end of 1941, the backlog was just under $34.5 million; the backlog at the 1942 year's end amounted to nearly $47 million, 59 percent of which were direct war orders.

In mid-1943, Union Switch & Signal reached its glory days with employees numbering 6,350, due to the intensive manpower requirements of the World War II production line. By 1944, the workforce dropped to 3,500 as production needs were satisfied. On November 3, 1944, a number of Union Switch & Signal employees attended a ceremony at the Swissvale plant as the company received the Army-Navy "E" Production Award in recognition of outstanding war production. Major General Charles P. Gross, chief of Transportation of the Army Service Forces presented the award to George Blackmore, who accepted it on behalf of the management and employees. W. H. Cadwallader, vice president and general manager, delivered the address.

Government contracts did not eliminate the need to continue their design and production of railway safety equipment during the War years. In fact, the contracts were handled in parallel with an increased output of the Company's regular line of railroad equipment required to keep the nation's railroads safe during the war. Design progress on the "Union" Coded Track Circuit Control continued in 1940, with the following first installs for:

• Cab signaling without wayside automatic signals in steam territory

• Approach-energized tuned alternators as the sole source of cab signal energy

• Single-track territory with Centralized Traffic Control (CTC) and

• Absolute Permissive Block with three- and four-indication signaling using coded detector track circuits and line wire signal controls.

Follow-up developments for the first installation of four- and five-indication signaling in electrified territory and three-indication signaling for either direction operation using polar reverse codes were completed by 1934.

According to a 1945 article entitled "Men and Women of Wartime Pittsburgh," the rail industry was better able to support the transportation requirements of the war effort due to a "better cooperation between railroads and shippers," higher capacity cars and faster operating speeds…"modern air brakes and signal systems play a vital role in safeguarding and expediting volume traffic."

Union Switch & Signal continued to support the rail industry requirements of the war in parallel with the immense workload of war materiel.

> **"Perhaps no industry is affected by the ups and downs of the economy so much as are the railroads."**
>
> — The Bulletin Index, Section II – April 29, 1943

Union Switch & Signal continued to develop, improve, redesign and add to their own cadre of products. Improved functionality in relays, rectifiers, searchlight signals and switch locks kept the men and women working around the clock in the engineering department as well as in the manufacturing plant. Of particular interest were the new M-22A dual control switch and lock movements, a switch machine that could be operated electrically or mechanically, the M21-A electric switch lock, and the Sl-7 Mechanical Time Switch Lock, which interlocked to set an approach signal at "stop" for a sufficient time before a hand-throw switch machine could be mechanically unlocked. Yard track indicators and an array of upgraded and higher performance relays rounded out the decade's progress in rail safety equipment design.

In a World War II era advertisement, Union Switch & Signal touted that they had coded track circuits in use in over 3,500 track miles on 18 railroads in the United States. In 1942, George Baughman, Norman Agnew, Porter Place and Fred Tegeler developed non-vital carrier equipment to remove the limitations of the control center's location. As its name implies, the equipment used a baseline or underlying signal that "carried" the different codes between the control center and wayside locations. The development added

Top: A Pennsylvania Railroad class N5c built in March of 1942, barrels along the Eastern Region. This class of caboose was made famous in the post-WWII period by The Lionel Company having made a model of one.

Above: Operator using ITC to communicate with a train by voice.

additional code line sections that the carrier transmitted over the first section nearest the control center. However, a given carrier signal could only transport a fixed number of codes; if more carrier signals could be employed, many more codes could be put to work. The following year, they developed a multi-carrier installation that increased capacity sufficient to pay for itself in three years. Patents for this remote control system were granted in Canada in 1944, and in France in 1944 and 1947.

In August of 1945, President Blackmore showed interest in the potential business of a new development in the company, the Inductive Train Communication (ITC) system. This system provided for instant voice communication between the engine and caboose of the moving train, between one train and another train on the same or adjacent tracks, and between trains and wayside offices. Vehicles on the track transmitted via the rail, while office transmissions were dispatched via rails and associated line wires. This system was installed successfully on a number of roads, but the largest and most important so far as traffic volume is concerned was the installation on the heavy division between Harrisburg and Pittsburgh on the Pennsylvania Railroad in 1944. (The ICC assigned channels for Railroad Radio service following a Federal Communications Commission hearing on the use of radio in railroad applications.)

George Blackmore was at this time showing a keen interest in some internal business as well. According to the April 17, 1945 meeting minutes of Union Switch & Signal, Robert H. Wood became the Company's General Manager, a position that Blackmore had held since 1922. At the same meeting, a gentleman named Albert N. (Nathaniel) Williams joined Westinghouse Air Brake as a director. The wheels of change were beginning to turn. Six months later, during the October 19, 1945, board meeting, Williams stated that he would like to work in an executive and managerial position for both Union Switch & Signal and Westinghouse Air Brake, "as long as it was for a term of not less than five years." Williams stepped out of the boardroom to allow the other directors to hold deliberations. When he returned, Blackmore was re-elected as Chairman of the Board for both companies, and Williams was elected President of both companies, effective January 1, 1946.

The Pennsylvania Railroad Class N5c Caboose, built in March 1942, used ITC technology to communicate with the locomotive, as indicated by the tubing on the roof.

Williams was an interesting mix of top executive and in-the-trenches railroad man. He had come to the Westinghouse companies from the Western Union Telegraph Company, where he had served as President since 1941 and as Chairman of the Board since 1945. But his roots were in engineering and railroading. Hailing from Denver, Colorado, Williams received his Mechanical Engineering degree from Yale University and worked on various railroads as a rodman, machinist apprentice, brakeman, second hand and section foreman. From 1917 to 1921, he worked as a construction engineer and operating superintendent of several petroleum industries in Oklahoma, Texas and Mexico. He returned to the railroad industry, working at the Midland Valley Railroad as assistant General Manager in 1921 and General Manager from 1922 to 1926. By 1940, he was President of the Lehigh Valley Railroad in New York.

Once seated in the President's office, Williams was also named Vice Chairman of both the Union Switch & Signal and the Westinghouse Air Brake Boards at the board meeting on January 1, 1946. Three months later, T. B. Clement became Executive Vice President; and Herbert A. May retained his position as Vice President. (May had been Vice President since at least 1945.) The constant "changing of the guards," as it were, could not have proffered much surprise when, on July 1, 1946, "Chairman Blackmore stated that the office of General Manager, now held by Mr. R. H. Wood, is not now deemed to be a necessary part of the

On the international scene, Union Switch & Signal's holdings in the Kyosan stock were seized by the Japanese Belligerent Property Custodian during World War II. Union Switch & Signal made interlocking machines for the Burma-Assam Railway, and "engineered a very large order for the Soviet Government under Lend-Lease auspices."

Following the war in 1947, Union Switch & Signal appointed Cobrasma (Companhia Brasileria de material Ferroviario) as their exclusive sales representative for Brazilian railroads. The same year, they signed an agreement with the French Company covering patent and manufacturing licenses for signaling and copper oxide rectifiers for manufacture in France, Switzerland, Spain and Portugal. The agreement also allowed the company to import product into Belgium in exchange for the Union Switch & Signal licensing of the French company's copper oxide and signal patents in North and South America. Union Switch & Signal also terminated their 1938 signal agreement with the Westinghouse Brake & Signal Company in London, effective in 1948.

John M. Pelikan

Mountain Man Outwitted Russian Police for Love

Every company needs a little spice in its history, if only to remember that companies really are made up of real people with real lives – far-fetched as some may seem! John M. Pelikan was a small-town boy who grew up in the foothills of the Appalachian Mountains, attending high school in Cambridge, Ohio. He studied Electrical Engineering at Carnegie Institute of Technology in 1923, and shortly there-after was accepted into Union Switch & Signal's two-year apprentice course. He worked in the signaling field from 1926 to 1928, returning to Swissvale as a signal system designer in the General Engineering Department.

Pelikan was sent to Russia in 1930 as the Supervisory Engi-neer for a Soviet government project involving the installation and commission of $150,000 worth of signal apparatus over a 125-kilometer section of the Moscow-Riga line running from Pokrovsko westward to Volokolamsk. While there, he fell in love and married Vanda Olga Brazlova, a girl whose father (an engineer) had been killed by the "Cheka," the Russian secret police force set up in 1917 by the Bolshevik government. "Cheka" was an acronym for Chrezvychainaya Komissiya, or, "Extraordinary Commission," which was reorganized in 1922 as the GPU, and later as the KGB.

But Pelikan encountered serious problems, it seems, when he tried to bring his new wife home to Pittsburgh, but the Russian government denied her a visa. A failed escape at the border landed them both in prison, after which he was coerced into becoming a GPU agent in return for his release. They ordered him to go to Moscow, but to avoid having to spy on Americans there, he claimed that his firm had ordered him to return to America. The GPU promised him a visa for his wife after one year if he would steal blueprints from Union Switch and Signal Company and Westinghouse Electric Company, as well as conduct economic class espionage in both plants. According to the interview Pelikan gave the Baltic bureau on the "Tribune" on his journey back home, "They promised to supply him with money through their agent in America to pay his engineer friends to steal important blueprints."

Pelikan said that Union Switch & Signal supplied him with out-of-date blueprints so that he could get his bride home, but the Baltic bureau also offered to help him, and Pelikan was once again introduced to border runners – ones he hoped would be more successful than the first ones he bribed. A letter was delivered to Mrs. Pelikan at the Savoy Hotel, and a few days later, a peasant appeared, saying he had brought salt pork, a rare luxury in Moscow. He slipped her a note from her husband with instructions to meet the next night at the Windau train station in Moscow.

Vanda Pelikan told the GPU on their daily rounds that she was going out to a party, then gave her "friendly GPU shadow" the slip and scooted off to the station. At a point near the fron-tier, she de-boarded the train to embark on a 100-mile sleigh journey to freedom. After another round of bribes to Soviet frontier guards, her two young escorts led her on foot the last 20 "versts" (13.3 miles) to the farm were her husband awaited her.

W. H. Cadwallader, assistant to president of the Union Switch and Signal Company, told the press that Pelikan was indeed an engineer at the firm and was en route from Russia, but denied the Pelikan had any dealings with espionage or the worthless blueprints that the European press said Union Switch & Signal provided him.

Pelikan continued to work at the Swissvale plant until 1941, when he was assigned to the Sales department, first in Chicago and then in New York. He was appointed Assistant District Manager in 1951. Returning to Swissvale in 1953, he became Manager of Sales Promotion and worked in foreign sales in Sweden, Turkey, Iran, Egypt, Brazil and Venezuela until his promotion to Vice President of Export Sales in 1955. Throughout his career, he was involved in the design and sale of Centralized Traffic Control (CTC) systems. He amassed 35 U.S. patents with the company, many of them for various forms of direct-wire traffic control systems. His last patent in 1963 covered a spark plug with an integrated induction coil.

John Pelikan ended his 33-year career with Union Switch & Signal in 1957 to establish two corporations, Actioncraft and Transcontrol, both headquartered in Long Island, New York. Transcontrol was an engineering and manufacturing company for, oddly enough, railway signal apparatus and signal systems. Many years later, Transcontrol would be acquired by Ansaldo STS and merged with Union Switch & Signal.

Herbert A. May

company's organization, and that arrangements have been made for Mr. Wood's duties to be henceforth accomplished through his appointment by the President as General Counsel. After some discussion, a motion made and seconded, it was unanimously resolved that in accordance with the provisions of Article 4, Section 1.a of the By-laws, the office of General Manager be and it is hereby abolished."

T. B. Clement

At some point, a company knows when it has the right people in the right place, and the reign of Union Switch & Signal remained stable for the remainder of the decade – with the slight change in Blackmore's title to Chairman of the Board and CEO for both companies. On October 2, 1948, however, George Blackmore died. The plant and offices suspended operations for five minutes in honor and respect to his memory.

An artist's depiction of "Tomorrow's Trains" in the 1947 Westinghouse Air Brake annual report.

Men and women assembling CTC equipment in the Swissvale plant.

10   Building a Conglomerate

How Union Switch & Signal managed to develop significant new technologies with the plethora of steel strikes, the growth of the company into a corporate conglomerate, and the cancellation of one project after another in a weakened rail industry is a bit of a mystery. Like tires on a car, basic components such as relays and switches were reliable, but required periodic replacement. It was those products that kept the company afloat during the lean times.

Three major steel strikes in the 1950s wreaked havoc on United States industries. President Harry S. Truman nationalized the American steel industry hours before workers walked out on April 9, 1952, but the steel companies sued and on June 2, the United States Supreme Court ruled that the President lacked the authority to seize the steel mills. The strike began and lasted 53 days. The steel strike of October 1957 noticeably held up production at Union Switch & Signal, but it was the 1959 strike that severely impacted the rail industry and its suppliers. Dozens of industries were forced to lay off workers as the 116-day strike – the longest steel strike in the nation's history – impacted Detroit's automakers, held up the Air Force's new Intercontinental Ballistic Missile launching base in Denver, and crushed the already hurting railroads. According to a Time magazine article, railroad employment on September 10 fell to 797,195, the lowest the industry had seen since 1900. The railroads

Women were trained in coil winding and other manufacturing skill sets during the war and continued to work after the war was over. This woman worked in the CTC plant in 1956.

lost an estimated $450 million in gross revenues during the 13-week strike. So damaging was the impact, that President Dwight Eisenhower invoked the Taft-Hartley Act in October and the Supreme Court upheld that the invocation was constitutional. Strikers grudgingly went back to work, but the impact of their strike carried over into the 1960s.

On the corporate home front, Albert N. Williams' five-year contract as President of Union Switch & Signal expired on December 31, 1950, but the Board determined "continued service of six months would be for the best interest of the company." At the December 11, 1950, Board meeting, Williams tendered his resignation effective January 15, 1951, after which he was unanimously elected Vice Chairman of the Board, effective January 15 to June 30, 1951. Williams then retired with a $16,000 a year pension and became mayor of Littleton, Colorado, from 1957 to 1959.

At that same meeting, Edward Boshell was elected Chairman of the Board and President of Union Switch & Signal effective January 15, 1951, with a salary of $25,000 per year. Herbert A. May, who joined Union Switch & Signal in 1936 as a Vice President, was designated Senior Vice President for both

George Washington Baughman

companies, a position which he held until his retirement in 1958. Several other vice presidents elected included Robert H. Wood as Vice President 1951, and as Vice President and General Counsel 1952 to 1967; Albert M. Wiggins, who entered the employ of Westinghouse Air Brake in 1933 as Vice President and General Manager 1951 to 1960; and George W. Baughman as Vice President of Engineering from 1951 to 1958.

On July 5, 1951, The Westinghouse Air Brake Company, Westinghouse Pacific Coast Brake Company and Union Switch & Signal Company merged, with Union Switch & Signal Company as the surviving company. Its name changed to Westinghouse Air Brake Company. The Company then operated as two divisions – the Air Brake Division and the Union Switch & Signal Division. So while Union Switch & Signal *appeared* to be a division of Westinghouse Air Brake Company (its former parent company), it was from this point on *actually a division of its prior corporate entity*. All activities of Union Switch & Signal Division were henceforth to be carried out under the supervision of a Vice President and General Manager. In the Westinghouse Air Brake 1951 Annual Report, Boshell stated, "Many economies and substantial savings in taxes will result from the new simplified structure." Suffice it to say, Uncle Sam did not agree with the Company's "new math plan," and a requested $187,771.82 excess profit tax and interest refund in 1951 was subsequently denied.

When Edward O. Boshell, 51, became President of Union Switch & Signal (aka Westinghouse Air Brake Company), he discovered the company was sitting on more than $25 million in "excess working capital." He said he would "expand and diversify the company's business

> **Union Switch & Signal applied the first commercial application of the transistor to cab signaling equipment. Albert Kasmir did much of the work to revise the cab signal equipment – eliminating the vacuum tube and replacing it with a transistor.**

Union Switch & Signal built the first supersonic flight simulators for the United States Air Force to train pilots on flying the F-100 Super Sabre jets.

as quickly as possible." In short order, he bought Melpar, Inc. of Falls Church, Virginia, the Le Roi Company of Milwaukee, Wisconsin, and the George E. Failing Supply Company. This expanded the Company's product offerings to include Melpar's F-100 Super Sabre cockpit simulators, which Union Switch & Signal manufactured in the early- to mid-1950s; Le Roi portable drilling rigs and related equipment for the water, oil and gas industries; and George E. Failing portable rigs for oil exploration. In 1955, Union Switch & Signal built the first supersonic flight simulator for the Unites States Air Force to train pilots on cockpit indicators and controls. The Company relocated the Le Roi Company to Sydney, Ohio, where they produced the Le Roi Tractair 125 into the 1960s, a tractor used mostly for construction and for running jackhammers and sandblasters.

Buying these companies drained most of Union Switch & Signal's cash coffers, but Boshell was not finished. In May 1953, he took out a loan with Pittsburgh's Mellon National Bank & Trust Company and other institutions to invest in 60 percent of R. G. LeTourneau Inc.'s plant assets in Peoria, Illinois, securing its widely popular earth-moving equipment line. Rumors of the deal skyrocketed LeTourneau stock from only 20 cents a share earlier in 1953 to $43 upon the announcement of the buy. Purchasing LeTourneau's major assets instead of its stock allowed Boshell to increase Union Switch & Signal's depreciation base by $18 million, giving the company the ability to write off half of it within ten years. In contrast, Robert LeTourneau, often referred to as "God's Businessman," dedicated 90 percent of his company's stock to the LeTourneau foundation, which sponsored Christian missions and financed LeTourneau Technical Institute, a school designed to educate World War

An earlier version of the 516 was this 1941 Style-C, 502A time code CTC machine from Palmer's Cove on the New York, New Haven, & Hartford Railroad. It controlled an interlocking at Milepost 128 between Groton and Mystic, Connecticut. From the private collection of Rich White II, of Reading, Pennsylvania.

Photo credit: Joanne L. Harris

Charles "Chick" B. Shields obtained the first patent in which the transistor was used for voltage regulation.

II veterans. All in all, Boshell's spending spree led to the company's conglomerate of the 1950s and 1960s.

The Board decided to separate the functions of Chairman and President, and on September 1, 1956, A. King McCord, formerly president of the Oliver Corporation in Chicago, assumed the duties of President and Chief Executive Officer. He remained President until the company was sold to American Standard in 1968. Boshell remained Chairman of the Board until he resigned in October 1957. Edwin Hodge Jr. replaced Boshell as Chairman of the Board.

Settling into the business at hand, Union Switch & Signal made headlines in 1952 when it installed the world's first Push Button Controlled Freight Classification Yard at the Conway Yard, 22 miles northwest of Pittsburgh in Conway, Pennsylvania. The yard had two separate yard facilities and 107 classification tracks, some of which were more than a mile in length. Conway Yard boasted every automatic "bell and whistle" of the day: electronic computers, radar, automatic switching, coded circuits, inductive trainphone, cab signals, electronic scales, micro-talkie radios, automatic floodlights, pneumatic tubes and tape recorders. A car retarder operator could push a single button on Union Switch & Signal's VELAC Automatic Classification Yard System to classify an entire train via electro-pneumatic retarders and switches. A computer collected data such as track rolling resistance, weight classification and track fullness to determine the correct exit speed from the retarders, then controlling the retarder to obtain that speed automatically. Union Switch & Signal's Paul N. Bossart, whose engineering career netted him more than 100 patents, secured the patent for the inductive train phone – a telephone communications system for trains using technology that prevented eavesdropping. (The research took nearly two decades to perfect.)

The CTC lab in Swissvale put systems such as this Model 516 through high integrity testing before they were put into the field.

The VELAC Automatic Switching Machine allowed the hump conductor to place destination storages into the relay banks of the automatic switching system either manually or automatically. This machine also fed the coded tape into the switching system and controlled the Inductive Cab Signal System. The new technology gave the Conway yard the capability to smoothly handle thousands of cars a day, rolling from two directions, and lining them to their proper tracks for their next journey. Another milestone came in 1953, when Union Switch & Signal installed the world's first Centralized Traffic Control (CTC) system with sole control of a rail system using microwave transmission for US Steel on a new line for the Orinoco Railroad in Venezuela.

Benjamin V. Becker was the longest serving Union Switch & Signal director. He was elected to the Board in 1913 and served continuously until 1952, a period of 38 years.

Some of the other significant installations that Union Switch & Signal made in the 1950s included the Union Pacific Railroad's coded cab signaling system with "whistle and acknowledgement" around 1950. The system was first installed as a two-indication system in the section from Green River to Laramie and as a three–indication system between Omaha and North Platte. Union Switch & Signal built several other three-indication, two-speed coded continuous speed control system for the Long Island Railroad; one used Type EL electrical equipment and another used a "529 indication system." The 529 indicated the positions or conditions of multiple functions on a number of wayside locations to one or more office or registry locations. It employed a two-wire polarized DC line circuit extending from the office or central location to the ends of the indicated territory with transmitters on the waysides and receivers at each office or registry location. Continuous cab signaling enabled the engineman to see the maximum speed limit below which he was permitted to run. The Company also installed four-indication, two-speed coded continuous speed control system for the Richmond, Fredericksburg and Potomac Railroad and the Central Railroad of New Jersey, and a Type "E" version for the diesel-electric locomotives of the Chicago, Burlington and Quincy Railroad.

In the early 1950s, Audio Frequency Overlay (AFO) technology was added onto track circuits, based on the teamwork of Chick Shields, Tacky Staples, Leslie R. Allison, Walter Quintin, Philip Luft and Herman Blosser. By 1954, AFO technology was used for switch lock release and in 1955 for highway crossing signals.

In 1955, Westinghouse Air Brake Company as a whole reached an all-time sales high of $172,502,277.

Union Switch & Signal continued to develop time-revered product lines that took them through the decade as well as some experimental products that did not thrive so well. Transistors found their way into the Audio Frequency Overlay track circuits and the Union Carrier telephone, which imposed a voice-modulated, high frequency overlay on line circuits for communication on the railroads as well as mining applications. The electro-pneumatic DA-10 switch machine reduced the total "throw" time from 0.6 to 0.4 seconds. The throw time was the amount of time for the rail to switch from one position to the other. This in turn cut the detector track circuit length ahead of the switch points from 22 to 14 feet. Finally, the 528 Code Control System provided an economical, high-speed method for remote control of a large number of interlockings over an extensive territory. It was a multiplex scheme consisting of two independent systems – a multi-station control system carrying data from the control machine to the interlocking machine and a single-station indication system to transmit data from the interlocking locations to the control machine. This system significantly increased the speed of data transmission.

Frank Himmler

The Bird Dog

Many people say they have "seen it all." Frank Himmler, 95 years old in 2011 and a former inspector at Union Switch & Signal, probably lived through more of the company's history than most other employees. Born in Cumberland, Maryland, he graduated from Perry High School in Pittsburgh in 1935, toward the end of the Great Depression.

"I had no money," Himmler recalled. "It was $150 a semester to go to Carnegie Tech. So I went to work and went to night school. When the war came, I dropped out of school and just worked. I started out making $30 a week at the Switch. That's when [my wife and I] decided to get married. We could afford to do it. We could pay for an apartment."

George Blackmore was president when Himmler started working on the 500-series time code units, primarily for centralized traffic control. The first time he worked overtime was Sunday, December 7, 1941. "My car radio broke on the way to work so I heard nothing of [the Pearl Harbor attack] until I got home. There were only two of us working on a control machine that day. We started working overtime seven days a week." Frank Himmler was initially the sole inspector in the Relay Control Group in Swissvale, until he trained 12 new men and became Supervisor of System Testing. After the war, he worked with engineer Frank Pascoe on a time code control machine for a new system spanning Laramie to Cheyenne, Wyoming. "At that time, I didn't quite relish sitting behind a desk. I felt I was doing more good in the field."

So back into the field he went. In the 1950s, Union Switch & Signal began implementing its time code systems in the arena of natural gas transport at

"Anything I ever saw, I just liked to know how it worked."
– Frank Himmler

Colombia Gulf Transport in Louisiana. Himmler was on the job. "The code systems were ideal for the pipeline. You had a control system and machine the same as you would on a railroad, except you were sending code steps out to tell a sequencing panel in the field to turn on and off compressors. Moving gas was like moving trains. The dispatcher knows where the trains are because the control panel lights up where the trains are. With pipelines it's the same thing – gas or oil coming through." Union Switch & Signal exited the pipeline business when the railroads decided that the pipelines were competition, Himmler said.

Himmler's career led him work on the Frankford "L" in Philadelphia, the US Steel mine railroad in Port Chartre (north of St. Lawrence in Canada), and on the track circuits at the Montreal Expo in 1965. After moving up to Engineering in 1965, he worked with Jim Winning, who was in charge of the subway section in New York. He became a Transit Engineer about 1970, and worked on the PATH project under the World Trade Center as resident engineer. Then it was off to Atlanta, Georgia's MARTA yard, before he retired at the end of January in 1981.

At a retirement dinner for Union Switch & Signal employees, Frank Himmler learned something new from Bill Dufer, a former colleague. "Dufer told me, 'When things got tough, they'd say it's time to get the "Bird Dog" out there.' I'd been retired four or five years before I ever heard the nickname. They thought of me as being the person who could "sniff out the trouble."

Crawford "Tacky" E. Staples

In 1953, a letter from Charles Wheeler (Union Switch & Signal Sales Manager) to Vice President and General Manager Wiggins confirmed that their 4-point DN-11 shelf relays were sold to Edgerton, Germeshausent and Grier (EG&G), a defense contractor for nuclear research. The letter stated: "It is our understanding that these relays are used on various projects which are more or less secret and that the company is presently engaged in work with the Atomic Energy Commission." The DN-11 line switch controllers were slightly modified as timing and firing signal relays to detonate atomic bombs during underground or surface testing. The Trinity test at Alamogordo Test Range in New Mexico was conducted with a six-mile, (and later a ten-mile) separation between the bomb and its control bunker, using radio control technology. "Trinity test" was the code name for the first test of an atomic bomb.

A February 1955 shareholder letter that stated, "Union Switch & Signal Division was also affected by the drastic reduction in capital expenditures by railroads," best describes the remainder of the decade. A number of significant projects were either cancelled or postponed, the Le Roi Division was down, as was the Air Brake Division that suffered an estimated $1,600,000 in flood damage at the Wilmerding plant. In spite of these difficulties, the production of aircraft flight simulators for Sabre Jet pilots continued at Union Switch & Signal, as well as participation in guided missile projects. Another first in 1955 was the design and production of a remote control system for an unattended compressor station on a gas pipeline.

The decade ended with Union Switch & Signal continuing support for the TALOS, ATLAS and JUPITER missile programs as well as the EXPLORER and VANGUARD satellite projects, supporting its sister company Melpar by building parts for these programs. CTC installations were now controlling more than 20,000 miles of railroad, along with signal systems for subways, and electronic computer-controlled classification yards. Despite all this progress, passenger traffic in the 1950s averaged less than 10 percent of freight revenue; by 1959 it had dropped to 8 percent, which led to railroads dropping passenger service. At least 11 Class 1 railroads no longer offered passenger service by 1959. (In the mid-1950s, there were 126 Class 1 railroads, defined as having a minimum $3 million in annual revenue.) Yet the railroads' hands were tied. They could not close a depot, stop running passenger trains or raise fares without federal government approval. The railroads were facing financial crisis, and the government finally came to the rescue. In 1958, the government passed The Transportation Act of 1958, which attempted to bolster the commercial railroads by granting the ICC money to loan to the railroads along with the authority to fix prices and allow a railroad to discontinue passenger train service. Within a few years, railroads received approval to drop almost 1,000 passenger trains.

A bronze bust of George Westinghouse was permanently enshrined on December 1, 1955, when he was elected to the Hall of Fame for Great Americans at the Bronx Community College at New York University.
Photo credit: Joanne L. Harris

Westinghouse Air Brake Company
Annual Report
1953

The annual report for Westinghouse Air Brake (which at this time was Union Switch & Signal under a new name) touted the many areas of the conglomerate built up by then President Edward Boshell.

Union Switch & Signal launched into outer space fame when its relays initiated lift-off for the lunar module to leave the moon's surface.

11   Transit Signaling Takes Center Stage

The Lackawanna Railroad sign can still be seen on the Hobeoken, New Jersey, terminal. *Photo credit: 6K Trammrunner 229*

The 1960s did not bring about the financial respite that the railroads so desperately needed. The ICC approved a number of major mergers in the 1960s for struggling railroads who saw them as a survival tactic. Mergers included the Delaware, Lackawanna and Western (Lackawanna) with the Erie line (renamed the Erie-Lackawanna); the Chesapeake & Ohio with the Baltimore & Ohio to create the C&O/B&O; and the Atlantic Coast Line and the Seaboard Air Line to become the Seaboard Coast Line. Soon after, the Norfolk & Western received approval to acquire the Wabash and the Nickel Plate lines, if they agreed to also acquire the Delaware & Hudson and the Erie-Lackawanna. Other mergers followed, the largest of which was the New York Central with the Pennsylvania Railroad, combining 21,000 miles of track to become the Penn Central. The Penn Central, saddled with disparate signal and computer systems from the start, was losing $1 million a day by June 1970, and went bankrupt. The domino effect was imminent for railroads that fed into the Penn Central. The Lehigh Valley Railroad, the Reading, the Erie-Lackawanna, and the Lehigh & Hudson River were all defunct before 1972 ended.

The demise of so many long-standing lines had a natural trickle-down effect on the industry's suppliers. The early 1960s found Westinghouse Air Brake's companies faring poorly in sales – again. The Le Roi business lost speed during a relocation of its production facilities, and Union Switch & Signal "incurred unusual costs during the year" in the midst of declining sales. Union Switch & Signal had acquired the industrial radio communication business of The Bendix Corporation and opened the new manufacturing plant in Batesburg, South Carolina, in January 1962 to produce radio equipment as well as miniature and micro-miniature electronic components. Contracts dribbled in, but the costs associated with integrating the acquisition into Union Switch & Signal's operation and with the establishment of the new Batesburg facility resulted in a greater decrease in profits.

The Batesburg, South Carolina, manufacturing plant ribbon cutting ceremony was attended by (left to right): Meredith Amick, executive director of the Batesburg-Leesville Chamber of Commerce; Francis C. Jones, state senator of Lexington County; Geoffrey A. Overton, WABCO plant manager; Uriah Collum, Vice President of the Twin city Industrial Corporation; Martin Rawls, mayor of Leesville; and A. C. "Red" Jones, mayor of Batesburg.

Edwin Hodge, Jr. was Chairman of the Board and A. King McCord was President of Westinghouse Air Brake when in 1965, Hodge asked to be relieved of his responsibilities as Chairman. McCord was elected to the office while remaining President. Albert M. Wiggins, Vice President and General Manager, retired on August 31, 1960, after 27 years of service and Robb W. James stepped into the position of General Manager, becoming Vice President in 1961. Jack C. Croft, Jr. was listed in the newly-formed position of Vice President Marketing. Robb James resigned in 1964, and was replaced by Union Switch & Signal's former Vice President – Manufacture and Purchasing, Charles D. Howell.

Government regulation once again impacted the railroads during the 1960s. Railroads were hauling about the same tonnage as they had 15 years prior; however, the trucking, barge and airline industries enjoyed a substantial increase in tonnage hauled, due in part to less stringent government regulation on non-rail modes of transport. Railroads thus curtailed major equipment purchases as they awaited government decisions on proposed mergers. It was a waiting game for everyone concerned.

The Westinghouse Air Brake Company Mass Transit Center was created in 1962 to target future light rail business opportunities, integrating the knowledge base of the Union Switch & Signal and Westinghouse Air Brake divisions, Melpar, and other Westinghouse Air Brake operating units that had specialty skills in the mass transit field. The Pittsburgh-based Center provided "information, advice and service to urban planners, engineering consultants, transit authorities and others interested in the problems of modern mass transportation." When San Francisco's proposed high-speed, computer-controlled transit system, the Bay Area Rapid Transit District (BART), voted to invest $792 million in 1962 for a three-county mass transit system, Union Switch & Signal was ready, willing and able. Union Switch & Signal submitted proposals for fully automated traffic control from a central computer, brakes and transit car trucks, all of which were accepted for test. The federally sponsored, 1964 Urban Mass Transit Act provided money to assist and encourage planning by area-wide urban authorities. President Lyndon B. Johnson, in his State of the Union message, lent support to the development and test of high-speed rail transportation between Boston and Washington, but the lines would have to be built before switch and signal suppliers would ever see a contract.

In 1962, U. S. Steel contracted Union Switch & Signal to build an Automatic Train Operation (ATO) system for driverless train operation from their iron mine in Atlantic City, Wyoming, to the Union Pacific interchange at Rock Springs, Wyoming. After the system failed to meet all of the expectations of the customer, Union Switch & Signal learned that their customer had no intention of running heavy ore trains over a line with long

Union Switch & Signal built remote control systems to control the flow of gas or oil, such as these gas pipelines in Hobbs, New Mexico, during the 1960s. *Photo courtesy of Frank Himmler*

What the astronauts remember most
is that we got them <u>off</u> the moon.

The relays used to initiate
lift-off of the lunar module
were one of the many uses
Union Switch & Signal found
for micro-miniature devices
for the aerospace industry
during the 1960s.

In early 1969, the company established an Aerospace Department, headquartered in the Swissvale plant, to "manage all phases of the miniature electrical relay business manufactured at the Batesburg plant." Inventor Ray Franke built the miniature, 26-volt, electrical relay switch that was used on Apollo 11's lunar landing module.

two percent grades without an engineer. The contract was more of a ruse to get the Wyoming legislature to repeal its five-man full crew law. A few years down the road, Canadian National also tried to implement ATO on a line, but it failed to come to fruition as well. Since then, no freight railroad has attempted to implement a driverless system on long distance or heavy-grade lines.

Concentrating on new technologies that would provide a near-term impact on the rail industry, Union Switch & Signal began manufacturing miniature and micro-miniature electronic components for missiles and aircraft fire control systems, including one model for Project Mercury, the nation's first manned space program, in 1962. A few years later, they adapted their rail technology to create a Centralized Transport Control for the oil and gas industry. An operator could monitor and control the flow of oil or gas along thousands of miles of pipeline through a series of meters, gauges and readout instruments. He could adjust the rate of flow and pressure, operate automatic compressor stations, start or stop engines, control engine speeds or even shut down a station if necessary, all from a single console. Electronically-coded information was sent to read-out instruments, which decoded and displayed the messages in alphanumeric form on control panels located in the locomotives. (Read-out instruments were also manufactured for the "P3V-1 Antisubmarine Warfare" program.)

On the international scene, Union Switch & Signal installed signaling systems in the Great St. Bernard Tunnel between Switzerland and Italy in 1964 through its European companies and their subsidiaries. Union Switch & Signal also secured a $1 million order from the Italian State Railways to provide highway crossing protection systems at 125 crossings.

119

Ray Franke

The Engineer Who Couldn't Quit

Ray Franke grew up in Millvale, a suburb of Pittsburgh. "Union Switch & Signal had a program in which they would choose several high school kids and offer them the opportunity to go to night school. "I wanted to go to college and be a schoolteacher, but they made $2500. Engineers made $5000. So I thought I'd become an engineer."

"A friend of mine's father worked at the Switch, and I found out about the apprenticeship program. There were four of us. His father vouched for me, they interviewed me and accepted me. While I was sitting in the lobby, I thought I might as well tour the plant. I walked through the buildings and into a restricted area – they were building Link Trainers at the time. The chief guard asked me what I was doing there and I told him I was taking a tour!"

Franke started in August of 1955 as a Laboratory Assistant in the Engineering Department, under Crawford Staples, making $220 a month. His group was responsible for making track circuit calculations, but he knew little about electricity. "I didn't really know what I was doing. I just followed the process. Tony Ehrlich was my de facto mentor. He took an interest in me because I took an interest in what we were doing." He later ventured into micro-miniature relays. "I worked under Max Adkins, a department head of General Apparatus, developing miniature relays and a pipeline bus. The pipeline business in some ways parallels the type of things that we do in Centralized Traffic Control – another natural outgrowth." Franke suggested to Adkins that he thought he could make an extremely small relay. "At the time, it may have been the world's smallest relay. They came to realize the potential of it. I never tried for a patent, but it ultimately made the Apollo program."

"I graduated in the spring of 1963 with a Bachelor of Science in Electrical Engineering. Terry Bollinger transferred me into Product Development working on radio control equipment. About that time, there was a big push for the BART project in San Francisco. That was the birth of, and the genesis of, rapid transit in the world." The group worked six to seven days a week, but lost the bid to General Electric. "We now had a product sitting in limbo until the World's Fair in '67."

Of Franke's 15 patents, some of which are joint patents, he is most proud of the MicroTrax track circuit. In the late 1970s, Union Switch & Signal launched into a new coded track circuit for mainline railroads. Ray Franke was a section manager for the project. Tony Erhlich constructed an elaborate mathematic model that closed the loop between theory and practice. "We also developed – which was unique at the time – the ability to adjust the track circuits simply on the basis that operator merely had to tell the processor the length of the track circuit. It would then adjust itself accordingly. That was the advent of MicroTrax. BNSF tried MicroTrax in an eight-mile-long tunnel in Washington State's Cascade Mountains and replaced several Electrocodes with one MicroTrax. That set the stage and it caught on. I had the idea, but Jim Brown and John Darrow did the hardware design, the two software engineers sweated blood getting it right."

Ray Franke was involved in the M-23 switch machine Electronic Circuit Controller (ECC) development and LED signals. "We used to laugh that we were called Union Switch & Signal but we had no signals – until the LED signal. I wrote a white paper on the 'Case for the Naked LED.' I proposed to put the active electronics (conditioning circuit), in a wayside case instead of at the signal head so you'd only need one conditioning circuit for all three lights. Larry Weber and Jim Werner got the patent because they made a lot of changes from my initial idea. It was nominated for the Ansaldo Innovation Award three years later."

Franke was still working part-time until March 2011, 54 years after he started. Why did he stay so long? "I've had interesting things to do. I've enjoyed my work. That's why I kept coming back."

George W. Baughman

The Man Content to Invent

George Washington Baughman (pronounced Boffman) graduated from Ohio State University in 1920 with a degree in Engineering. He went to work for Union Switch & Signal sometime in the early 1920s, and filed his first patent for an AC electrical relay in October of 1924. He is credited with the development of Centralized Traffic Control, or CTC, which greatly improved the safety, capacity, speed and economy of train movement. Baughman was a prolific inventor. While sources vary, he was awarded an estimated 111 U.S. patents between 1924 and 1966 (for which he received $50 per patent), filed under Westinghouse Air Brake as well as under Union Switch & Signal.

Baughman utilized the multiple lines normally used for telephone wiring to develop a system permitting simultaneous use of two line wires for telegraph, telephone and CTC transmissions. Filters installed at each field station allowed the transmission of an electric pulse or "conversation" over the same line without interference, saving immensely on copper wire. By 1941, Baughman refined his system for the multiple-line circuit and developed the 506 series of code systems in 1942.

George Baughman served as Assistant Chief Engineer, Chief Engineer and Vice President of Engineering for Union Switch & Signal. On December 1, 1955, he demonstrated Union Switch & Signal's crewless technology on a New York, New Haven & Hartford Railroad transit car with 75 guests aboard. Several years later, he penned an article for the 1962 May-June issue of Financial Analysts Journal, entitled "Step by Step Approach to the Crewless Train," by G. W. Baughman.

According to his son, George W. Baughman III, George Baughman also did relay research for the Manhattan Project via Union Switch & Signal, working in Columbus, Ohio. "His contribution allowed the pilot [Paul Tibbets flying the Enola Gay] to drop the [atomic] bomb without it going off until the plane was purportedly safely out of the way," Baughman III said. "They had done experimental work with much smaller bombs, but that was like going from a go-cart to a Maserati!"

After he retired in 1965, Baughman received the Elmer A. Sperry award in 1971 for his advances in transportation technology. When he retired as Vice President of Engineering in 1965, he was making about $27,000 per year. Ohio State University, Baughman's alma mater, bestowed upon him the first honorary degree of Doctor of Engineering, one of its highest honors, at their 1985 autumn commencement ceremony. "Unbelievable," said Baughman, 85, in a December 1985 interview with The Orlando Sentinel newspaper. "There must be 20,000 engineer graduates from Ohio State, living engineers, and the honorary degree goes to just one person. Unbelievable."

121

Not every development is a winner. In 1966, Union Switch & Signal developed an Automatic Car Identification (ACI) system – a laser-based optical system with black and white barcodes that electronically identified complete trains and individual freight cars during full speed operation. It was a pioneering effort, but the AAR chose Sylvania's "KarTrak" with color bar codes as the industry standard. From a sour grapes point of view, the rail industry only installed KarTrak during the late 1960s and early 1970s, until reliability issues prompted the Association of American Railroads (AAR) to stop requiring optical ACI labels.

Part of the proposed Automatic Train Operation (ATO) system for BART system was sold for installation on the transit system of the Montreal World's Fair (Expo '67) in Quebec, Canada. The Montreal project became the first urban transit system in the Western Hemisphere to implement the driverless cab-signal and speed control system. The City of Montreal needed transportation for the more than 50 million visitors and representatives of the 62 participating nations. They chose the Expo Express rapid transit system to travel through five stations along a 5.7 kilometer (3.5 mile) route. Union Switch & Signal installed its new ATO system with audio frequency track circuits on a fleet of 48 Hawker Siddeley H-Series variant cars configured in eight trainsets of six cars. Trains ran every five minutes and carried 1,000 passengers each. Over the course of the exhibition, the Expo Express transported 44 million passengers.

Union Switch & Signal provided Automatic Train Control for the 1965 Montreal Expo. The Expo kept conductors on the trains to make it appear they were not driverless.

The fair did not publicize that the trains were unmanned, as the organizers were concerned that visitors might not use the train if they

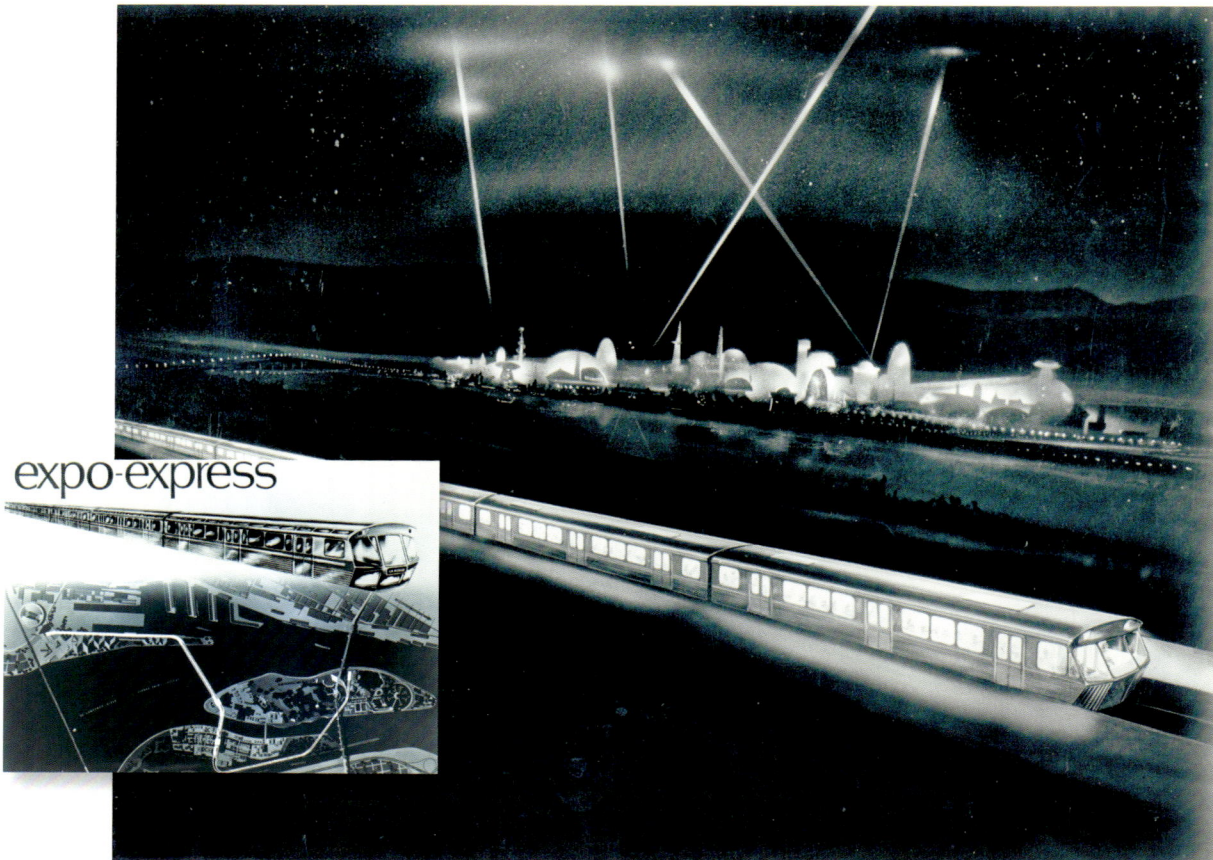

knew it was computer controlled! They even placed "operators" in the cabs and had them open and close the doors of the train to avoid visitor apprehension. The perceived "conductor element" offered one known faux pas during the entire fair, at La Ronde station. The conductor pressed the button to close the doors and proceed, but the computer automatically delayed the train to allow an oncoming train into the station. Meanwhile, the driver realized he had forgotten his lunch, but he could not exit the train as the computer put his train on "stand-by," thus keeping the doors closed. He crawled through the cab window to obtain his lunch. By the time he returned, his train was long gone. It crossed the bridge over the channel, proceeded along the seaway, and stopped at Ile Notre Dame station where an Expo official was waiting to crawl in through the cab window to open the doors for the unwitting passengers. After Expo '67, the City of Montreal bought the cars and offered service for four more years on a shortened route. In 1972, the trains were mothballed and stored on Ile Notre Dame. The project was to become the last real highlight for Union Switch & Signal in the 1960s.

The annual reports of Union Switch & Signal throughout the 1960s talked of poor earnings and struggles. Every year painted a hopeful picture that times would turn around, but sales were down and even if they had the contracts, there was a "shortage of engineers and skilled and unskilled labor" that tended to "restrain major increases in billings."

Frank Himmler inspected equipment from a precarious position in a remote Canadian location, about 200 miles from Sept Iles, Quebec in 1960. *Photo courtesy of Frank Himmler*

Something had to give. Robb W. James, Vice President and General Manager of Union Switch & Signal from 1961 to 1963, wrote in 1963 that the immediate outlook for sales of equipment was less favorable as rail spending continued to go towards rolling stock. In 1965, Westinghouse Air Brake Company President A. King McCord stated, "Traditional hardware, particularly in mass transit, is being replaced by new and smaller electronic devices…Rapid change means obsolescent inventory, and, in our situation, an old and over-sized plant." Times were looking gloomy for the conglomerate. The following year, Lawrence E. Walkey, president of Westinghouse Air Brake Division in 1966, echoed James' sentiment.

Orders at the Signal & Communications Division (as Union Switch & Signal was then called) increased 94 percent, but it was too little, too late. Westinghouse Air Brake was in trouble, and one could smell trouble in the air. This was an era of corporate takeovers "mushrooming" throughout the country, as the United States Court of Appeals defined it. The takeovers became so serious that Congress and the Securities and Exchange Commission (SEC) began to closely follow the flood of lawsuits that paralleled the buyouts.

Thomas Mellon Evans, a corporate entrepreneur and chairman of the plumbing industry's Crane Company, proposed a merger to Westinghouse Air Brake in May 1967, and began making substantial purchases of Air Brake stock the following month. An independent consulting firm gave Westinghouse Air Brake a "thumbs down" to the idea, and in November 1967, they declined further merger discussions with Crane Company. By then, Crane Company owned nearly 10 percent of Westinghouse Air Brake's outstanding stock,

World's First ATO

In 1962, Union Switch & Signal and General Railway Signal worked together to build the world's first Automatic Train Operation on the New York City's 42 Street – Grand Central to Times Square Shuttle. The project was a success, but it was only in revenue service for two years, from 1962 to 1964, due to a major train fire that ceased the shuttle operation. The two companies installed double and single-rail coded AC track circuits. The train doors only opened when the train was properly stopped, the gap filler (an extender between the platform and train) was extended and in position. The train safely carried passengers at a top speed of 30 miles per hour, and stopped within two feet of its intended stop.

New York City Transit Authority Chairman Charles L. Patterson gave the following remarks at the opening ceremony. "The four and a half minute route trip we are about to take on this first automated subway train is an end and a beginning...It is the end of the nearly three years of thinking about, developing, testing, installing, testing and retesting, and improving the equipment that control the three cars we are about to ride...In years to come this train and its control system will become as outdated as the Model 'T' Ford is today...My remarks on the continuity of transit progress are intended to emphasize the great accomplishment of the men and the organizations that made today's forward step possible...They are too numerous to be named individually, but without the ingenious and devoted work of the development and test engineers, and the large investment of talent and money made by the signal companies and other suppliers, this train could not have been automated...Great credit is due all of the men involved in this enterprise."

and continued to purchase more. Meanwhile, American Standard representatives proposed a merger to A. King McCord, and Westinghouse Air Brake directors agreed to it. American Standard, a competitor of Crane Company, manufactured household plumbing fixtures, air conditioning equipment, building products and transportation equipment, and they saw Westinghouse Air Brake as a neat fit into their transportation division.

Crane Company would not give up. It continued to accumulate Westinghouse Air Brake stock, holding about 32 percent by May 1968. The merger between Westinghouse Air Brake and American Standard became effective June 7, 1968, and under threat of legal action by American Standard under antitrust laws, Crane Company sold all but 10,000 of its shares of American Standard, and later disposed of all but 1,000 shares. The Crane Company appealed the decision in March 1969 to the United States Court of Appeals Second Circuit, claiming that American Standard manipulated stock and committed fraud in the purchase and sale of securities, and that Westinghouse Air Brake Company misrepresented American Standard's earnings and falsely stated in its proxy statement that "careful consideration had been given to the merger." The Court found that both parties had "carefully thought out, weighed and planned each step," and therefore, there were no improper motives and that no federal statute or law had been broken.

Westinghouse Air Brake's business was to be continued "thereafter by the present personnel of Westinghouse Air Brake in a new wholly-owned subsidiary of American-Standard which will be named Westinghouse Air Brake Company." American Standard quickly folded Westinghouse Air Brake Company into the Railway Products Group as the Westinghouse Air Brake Railway Division, and placed Charles Benjamin "Ben" Ramsdell in the combined role of Vice President and General Manager until 1973.

Opposite: Union Switch & Signal has maintained social activities for employees since George Westinghouse began a program in the company's infancy. The Union Switch & Signal bowling league began in 1919 and ran through 1985 in Swissvale. Notable league members included executives Albert Wiggins, Herbert May and R. H. Wood. The Batesburg, South Carolina, plant has organized softball leagues and continues to participate in the community's annual Poultry Festival and Parade.

Heimbuecher

Frances

Van Tassel

Zamaria

Hart. Capt.

GIRAFFES
1928 — 1929

Forty-first Annual 🎳 Bowling Banquet

FIRST YEAR FOR TEN PINS
1959 - 1960 SEASON
THURSDAY , MAY 5
CHURCHILL VALLEY COUNTRY CLUB

Bowling League leaders of 1928 – 1929

125

Aerial view of Swissvale Plant, Pennsylvania, circa 1970s

12   Rescuing the Industry

The 1970s brought more mergers, bankruptcies and inflation. Consumer costs nearly tripled between 1960 and 1980. The Common Market countries opposed most of the United States' proposals on trade and international monetary reform and maintained levies against American farm products. OPEC (Organization of Petroleum Exporting Countries) emerged, and soon France, Britain, Italy and West Germany each made private deals to buy Arab oil while the United States and other Allies were cut off by OPEC's embargo of October 1973. Diesel fuel jumped from 11 cents a gallon in 1970 to 81 cents in 1980. The oil and energy crisis that ensued had cars lined up for hours at gas stations in the early to mid-1970s.

Union Station, in Washington, D.C., serves the greater D.C. area with services from Amtrak with a complete redesign in 1988.
Photo credit: Matt H. Wade

Rail equipment costs rose nearly 2.5 times between 1970 and 1980. Railroad management faced severe labor cuts, even though they were now laying continuous welded rail to reduce maintenance cost, and installing computerized equipment to increase efficiency on the trains, rail, waysides and classification yards. The number of intercity rail transport running by 1970 had dwindled to 500 trains running on 49,000 miles of track. The railroads were running out of time and money, and passengers began to complain.

In October 1970, President Richard M. Nixon came to the rescue when he signed the Rail Passenger Service Act of 1970. The Act created the National Railroad Passenger Corporation (NRPC), which took over intercity rail passenger service. The new "Amtrak" ("America" and "Track") service opened for business on May 1, 1971, running a 21,000-mile network from Miami to Seattle and from San Diego to Boston. The remaining 13 Class 1 railroads paid a combined one-time payment to the NRPC of $190 million in exchange for permission to drop their passenger service. (The Denver & Rio Grande, the Rock Island, and the Southern kept their service to avoid paying the fee.) The hodgepodge of donated, aged trains and management's lax attention to train schedules sent the accounting books reeling into red ink within its first year.

Even though the government eased the pressure on railroads, freight business continued to decline. In June, the giant Penn Central declared bankruptcy, setting the record as the largest corporate bankruptcy in

The Rock Island E-6 Number 652 on the Midland Railway in Baldwin City, Kansas.
Photo credit: Quatro Valvole

United States history up to that time. Union Switch & Signal had a backlog of $50M in orders, inventory and work-in-process for the railroad. Heavy layoffs at the Swissvale plant were eminent. The rail industry's rate of return on investment slowly spiraled down from an average 4.1 percent in the 1940 to only 2.0 percent in the 1970s. The disappearing income translated into reduced speeds on over 47,000 miles of track due to unsafe conditions and billions of dollars in deferred maintenance. "Standing derailment" – in which standing railcars simply fell off poorly maintained track – became a common phenomenon.

Suppliers sought to bring solutions to the ailing railroads. In 1970, Union Switch & Signal developed the first digital classification yard control system for the Argentine Yard in Kansas City, Kansas. By the late 1970s, 26 of the more than 130 active retarder controlled hump yards in North America would be computerized or would be entering the process. Railroad yards such as the Louisville & Nashville at Louisville, Kentucky, the Seaboard Coastline yards at Waycross, Georgia, and Southern Pacific's West Colton Yard in Colton, California, installed highly advanced computer systems. The Southern Railway (now Norfolk Southern) installed a PDP-11 computer at Atlanta's 12-track Inman Yard that provided fixed exit speeds for routing and retarding. At Southern Pacific's Eugene Yard in Eugene, Oregon, a single computer at a control console used "dial-a-speed" to control retarders and relay-type automatic routing.

On September 1, 1972, The Signal and Communications Division of Westinghouse Air Brake Company returned to its prior name, "Union Switch & Signal Division."

Union Switch & Signal was still under the leadership of Charles Benjamin "Ben" Ramsdell as Vice President and General Manager. Ramsdell hailed from Washington, graduating from Catholic University in 1940 with a degree in Electrical Engineering. After a 23-year career with General Electric, he joined

Tower view of the Chicago, Milwaukee, Saint Paul & Pacific Railroad hump yard.

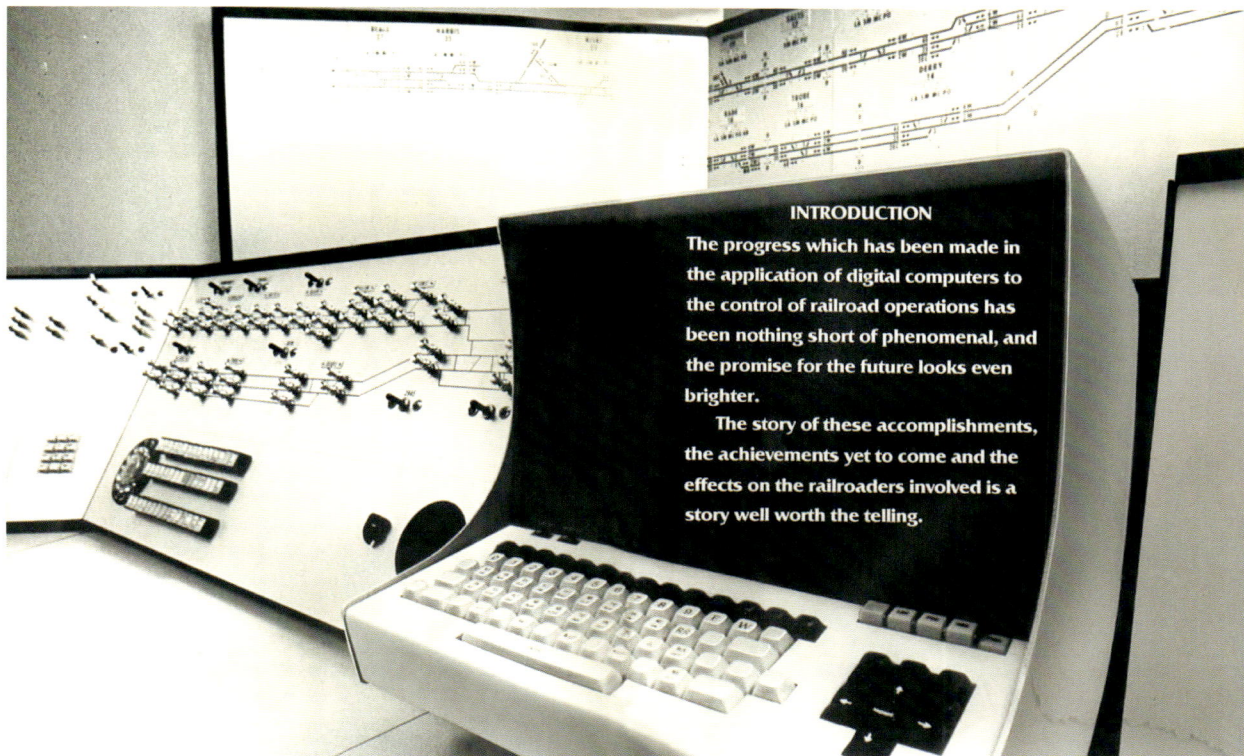

INTRODUCTION

The progress which has been made in the application of digital computers to the control of railroad operations has been nothing short of phenomenal, and the promise for the future looks even brighter.

The story of these accomplishments, the achievements yet to come and the effects on the railroaders involved is a story well worth the telling.

A 1976 Union Switch & Signal bulletin, "Computers vs. Relays" describes the progress made in the application of digital technology for railroad operations.

Westinghouse Air Brake in March of 1964 and within four years he was leading the company. Union Switch & Signal now had operating companies in ten nations and was selling products in most European countries.

The advent of the computer in the 1970s would change the railroad equipment industry forever. A 1976 Union Switch & Signal bulletin, The Progress and The Promise, described the needs for new types of engineers – computer designers, programmers and qualified real-time process control programmers. With 1,600 employees, the company continued to manufacture the "bread and butter products" that saw them through rough times when large systems were not marketable. Those staples included cantilevers for grade crossing warning signals, crossing bells, grade crossing signs and color light dwarf signals. Reinventing a bit of the past, Union Switch & Signal drilled their own on-site gas well at Swissvale plant to keep production flowing independent of outside sources. George Westinghouse would have been proud!

Glenn Edward Stinson became Vice President of Engineering in

"Tacky Staples" and Glenn Edward Stinson in the mid-1970s.

1970, working for Ben Ramsdell. Where many leaders came in with fancy degrees and management training, Stinson had started as low as you can get – digging ditches for signaling systems in Oklahoma in 1947. He worked his way up from system installations to maintaining signals. In time, he became a system designer in the general offices of the Rock Island Railroad. He spent one year at Southwestern State College in Oklahoma, but left school shortly after he started a family. Stinson entered Union Switch & Signal's large Engineering Department as an Engineer B making $515 per month in 1956, and later moved to the company's design office in Chicago. He transferred as the District Manager when the Sales Engineering department reopened its St. Louis, Missouri, office. While there, he covered the Southwest region. Stinson relocated to Swissvale in 1969 as the Marketing Manager.

Coming back from the sales side into the engineering department, he objectively saw some of the problems that were affecting the business. "When I was in Sales, the General Manager would ask us what the company was doing wrong. This was in the 50s and 60s. I told them 'You never stay with something, you have no commitment to search out something and see it through.' The company needed discipline. People had to be accountable. Marketing would forecast, engineering would build up budgets, and then the forecast never happened and you had a big cost. That's what I inherited."

His first task as an executive was to trim the staff within the department. It was not easy for him, as he had never fired anyone before. The times were difficult, and there were stories of men who committed suicide upon losing their jobs. There were 335 engineers in the department in 1970. Within a year, the number dropped to only 185. "Engineering led the cost cutting program throughout the division. We were trying to do too much and thus, not doing well."

In 1973, American Standard dismissed the head of Westinghouse Air Brake's Europe division, and appointed Keith Bunnel to fill the position in Brussels. Ben Ramsdell took Bunnel's position as Group Vice President, Railway Products and Glenn Stinson was appointed Vice President and General Manager of Union Switch & Signal in September 1973. Union Switch & Signal and General Railway Signal were faced with growing global competition. Union Switch & Signal needed to downsize and reassess its product lines. Stinson wrote in a letter to the employees, "Our high-cost, large facility in Swissvale is a blessing if we have

Hot bearing detectors, such as this one installed in Naperville, Illinois, circa 1962, identified overheated bearings before they caused significant damage. The product worked, but was short-lived. *Photo credit: Frank Himmler*

Glenn Edward Stinson

He Walked Through the Shops

Glenn Stinson worked in a myriad of positions before he came into the job as Vice President and General Manager for Union Switch & Signal. Stinson knew what it took to work in the field, in Engineering and in Sales and Marketing. Working literally from the ground up prepared him for the ultimate task – bringing the company back to profitability.

While most executives obtained an undergraduate degree before attempting further education, Stinson took a unique route. When he became Vice President of Engineering in 1970, Union Switch & Signal enrolled 25 established managers in a charter Executive MBA program at the University of Pittsburgh. They looked for managers who had the street smarts and the intelligence to accomplish greater things. Glenn Stinson had such a resume.

> **"The key to the Switch is the engineers."**
> – Glenn E. Stinson

When Stinson was in Marketing, he discovered the Engineering department could not talk to the Italians. They came to Stinson and his marketing group, and together they developed peer relationships with their Italian counterparts. "It created a whole new atmosphere of cooperation." Stinson said. When he was working in Sales, he invented the sales budget and strategic planning. "I didn't know anything about it," he reminisced, "but it made sense to plan ahead."

At one point late in his tenure at Union Switch & Signal, Stinson had a boss who complained that he didn't spend enough time with his "higher ups". His bold but honest reply was simply, "The people above you don't create the business. The business is below you and with the customer." His 'Management by Walking Around' was admired by some and a bit feared by others, but he knew what he wanted to see from each department and he wasted no time finding the waste in the company. He reorganized Engineering, got rid of product lines that were draining the business, and asked everyone to just "work a little harder" to make the company succeed. "We strategically bid the big systems jobs to balance the revenue jobs. The big jobs make or break you."

Roy Stecker, an engineer who still works at the company, recalled, "Glenn would park out back by the shops, and he walked through the shops every day going into work." Another employee who recalled Stinson from his early days on the job said, "He would walk down the aisles with his hands behind his back. It was a little intimidating." To that, Stinson replied, "I was tough, but I was fair."

The day Glenn Stinson went to Washington, D. C. to sign the Northeast Corridor project contract, he penned the contract and then crossed the street to meet with, and agree to go to work for Illinois Central Railroad's (ICR) CEO Bill Taylor. Stinson left "the Switch" and put a plant in Beijing to manufacture brake shoes for the Abex Railroad Products group – part of ICR in Chicago. Stinson later bought out the railroad parts group and named it ABC Rail. He said he had his sights set on buying Union Switch & Signal at one point, but Ansaldo STS proffered a higher bid. He now enjoys his retirement with his wife in Ligonier, Pennsylvania.

Sedgwick N. Wight (in memoriam) of General Railway Signal, and George W. Baughman of Union Switch & Signal were the 1971 recipients of the Elmer A. Sperry Award for Centralized Traffic Control. Citations were also awarded to Clarence S. Snavely and Herbert A. Wallace of the Signal & Communications Division of the Westinghouse Air Brake, for their development of Centralized Traffic Control on railways. The award was given for "a distinguished engineering contribution, which, through application, proved in actual service, have advanced the art of transportation whether by land, sea or air."

volume but an albatross around our necks if we don't. I have reluctantly had to move products to other low cost facilities because we simply could not compete on those products from Swissvale."

Stinson recalled the problems of the company as this. "We developed card systems that worked but weren't accepted, remote locomotive controls, but didn't set up a marketing department to sell it. Hot bearing detectors, Identra, the early car identification system, never went anywhere. It was inductive – not the right technology. Inductive technology was used in transit systems until automatic car identification came along, but then computers made that obsolete. We created organizations that could solve real problems. Instead of doing everything everyone wanted, I'd challenge the research department to create something that would turn to gold. The research department was challenged to develop things that would sell. Every month of every year for seven years that I was General Manager, we made money." One of the first lines Stinson sold off was the company's two-way radio business. Motorola was their main competitor and Union Switch & Signal only had the small segment of railroad radios. It would take too much investment to make the line profitable, so it went…by the wayside.

The regional railroads got another boost when President Nixon signed another bill in 1974 to help keep them float. The Regional Rail Reorganization Act of 1973 designated $2 billion for track, right-of-way and equipment modernization. It also provided for the reorganization of railroads; the establishment of the U. S. Railway Association to replace the ICC; the creation of Consolidated Rail Corporation (Conrail); and financial assistance for state, regional and local transportation authorities to continue local rail service. President Gerald Ford signed the Railroad Revitalization and Regulatory Reform Act of 1976 that created Conrail from six bankrupt Northeast railroads. It also included reforms designed to make the rail regulatory system more responsive to changing circumstances. Conrail spent over $2 billion in federal aid, along with millions in private investments, finally seeing a profit in 1981.

This financial stimulation stabilized the market for suppliers, although 1976 was still a depressed year. Business was stable domestically, recalled Stinson, but only due to federal investment such as "Conrail's 'catch-up' modernization program." The following year improved Union Switch & Signal's bottom line, but the difficult economy and increased worldwide competition would be a challenge. In 1977, half of the company's income was a direct result of international sales, including a major signal modernization and rail traffic control project in Fepasa, Brazil, and the Netherland's Kijfhoek computerized classification yard.

The advent of the computer was both a blessing and a burden to the rail suppliers, who lost an intrinsic market advantage as the industry transitioned to microprocessor control. The venerable vital relays were difficult for industry newcomers to copy and certify, but microprocessors and their associated electronics were easy to procure and apply. The four leaders in the switch and signaling business would soon face a glut of new competitors in the marketplace. The new technology's ability to churn out a diverse array of "smaller, faster, cheaper" innovations would create more products from the 1970s to the present time than the industry had seen in it first 100 years!

The Audio Frequency track circuits use suitcase-sized "impedance bonds" to isolate the sensitive track circuit signals from the powerful current that feeds the trains electric traction motors.

Stateside, Union Switch & Signal secured an order from the Massachusetts Bay Transportation Authority (MBTA) to design, manufacture, install and demonstrate a method of Automatic Train Operation (ATO) for transit systems in 1977, which became the first analog Audio-Frequency (AF) Track Circuit in revenue service. Internationally, the company built a new plant in the suburb of Torino (Fiat town) and grew a $150 million dollar business there. Stinson was Chairman of the company at the time. He saw many opportunities in Brazil, France and Italy, and wanted Union Switch & Signal to be an international company. "We set out to do it, but American Standard didn't want us to go global. It took four years to secure the $15 million Ireland contract to automate the commuter lines in and around Dublin. We put a plant in Tralee – in the West. We could hire Irish electrical engineers from Trinity College Dublin for something like $12,000 a year and we could do the system engineering."

Nearly 11 years after the purchase of the Westinghouse Air Brake stock, Westinghouse Air Brake Company was merged into American Standard on December 31, 1978. Both Union Switch & Signal and Westinghouse Air Brake Company then became separate divisions of American Standard. Union Switch & Signal went forward with business as usual. They developed the first microprocessor-based railroad device, the DDL-601A Non-Vital Digital Link in 1979, secured orders for new Baltimore Transit System and won a contract to modernize signaling systems for the Algerian and Tunisian railways. Their Italian operation introduced the world's first solar powered highway crossing system for Saudi Arabia. Backlogs were at record levels. Life was good, and the biggest job was yet to come.

In 1976, Amtrak acquired the defunct Penn Central and New Haven lines from Washington D.C. to Boston on the Northeast Corridor, designating the 456 miles of multi-track right-of-way as passenger train-only traffic. Glenn Stinson met with the Federal Railway Association at that time. "I encouraged [them] to negotiate the project instead of bidding it out," Stinson recalled. "It would save them time and money over a full bid, and only two companies – GRS and the Switch – could handle the size of the project. In 1980, it took a long time. We had to deal with the consultants on the project. The Feds fronted the money for a high-speed railroad but they didn't know how to make it happen. They just needed a dedicated railroad.

Two Penn Central General Motors SD40s and a SD45 locomotive hauled loaded ore westbound at Duncannon, Pennsylvania, in August 1971.

Photo credit: © Paul Carpenito

By 1976 the number of Class 1 railroads dropped from 126 (each with at least $3 million in revenue) the mid-1950s to only 52, with the new requisite $10 million minimum revenue. A further change in definition to $88 million brought the number of Class 1 railroads down to 16. In 1995, there were only 11. Today, there are seven Class 1 railroads in the United States and Canada.

There were a lot of problems with specifications and interoperability."

Stinson's efforts paid off. Union Switch & Signal won a contract for a $45.6 million segment of the $100 million Northeast Corridor Improvement Project in 1979, signifying the largest signaling contract that Union Switch & Signal won during its first 100 years of operation. LeRoy "Roy" H. Stecker, who joined the company in 1966 as a circuit design engineer for the Wayside Engineering Group, served as Manager of Engineering for the new project. The project included the manufacture and installation of hundreds of relays, "M"-style switch machines, interlockings, position and color light signals, and the wayside portion of the cab signaling. The carborne cab signaling was already installed. "The ADL-256 discrete component was a non-vital product used on the North Corridor," Stecker said. "That

was the predecessor for Genisys®, a microprocessor-based, non-vital logic emulator, which replicated what non-vital mechanical relays used to do. Genisys was the predecessor of the vital MicroLok®." Once Union Switch & Signal installed the signaling, they had to upgrade the existing track circuits from 25 to 60 Hz, as the rail had not been electrified up to that point. The Northeast Corridor Improvement Project would take several years to complete, supplying Union Switch & Signal with steady work for the years to come.

In 1978 and 1979, Kenneth Liatsos and Arthur W. Ticknor were listed as Vice Presidents under American Standard's Westinghouse Air Brake Transportation Products group. Both men would each take the helm briefly as President of Union Switch & Signal.

Glenn Stinson, Vice President and General Manager of Union Switch & Signal, signed the Northeast Corridor (NEC) contract with NEC consultants in Washington, D. C. in February, 1979.

135

Nearly 55,000 employees of the Westinghouse corporations funded the Westinghouse Memorial in Pittsburgh's Schenley Park, which was dedicated in 1940 to honor inventor and industrialist George Westinghouse and his six major areas of achievement. On the far right is the plaque for Union Switch & Signal.

Photo credit: Joanne L. Harris

13   The Centennial, the Strike and the Sale

The economic downturn in the 1970s continued into the 1980s. The United States recession worsened, affecting nearly all the rail transportation markets. Railroads faced intense competition for freight traffic, but excessive regulation prevented them from competing effectively. Railroads needed a reprieve from regulation that would allow them to be competitive with other shipping industries, or they would not survive.

Congress had the option of nationalizing the railroads at an exorbitant expense (and with no guarantee that they could do a etter job of running them), or creating a more equitable regulatory scheme. Representative Harley O. Staggers of West Virginia sponsored a bill to deregulate the railroad industry while retaining the authority to protect shippers against potential price gouging. The Act would empower the railroads to decide what routes to use, what services to offer and what prices to charge. They could base rates on market demand, enter into confidential contracts with their shippers without Interstate Commerce Commission (ICC) approval, and expedite aban-donment and merger proceedings through the ICC. The Senate and Congress approved the Act and President Jimmy Carter, the peanut farmer from Georgia, signed the "Staggers Act" into law on October 14, 1980.

Southbound freight train hauling iron ore through Iron Junction, Minnesota, on the Duluth, Missabe and Iron Range Railroad.
Photo credit: © Wade H. Massie

Before the Staggers Act, railroads had the lowest earnings of any form of transportation. In the years that followed deregulation, they slowly began to recover. Within a few years, American railroads were in-creasing their freight traffic and making a healthy profit while in some cases lowering their rates. Between 1990 and 2009, return on investment averaged nearly 8 percent for railroads, up from a 2 percent average in the 1970s, and railroads invested over $6 billion a year in roadway, infrastructure and equipment improve-ments.

The railroads set a new course, one that established incentives for modernizing office, cab and wayside equipment to obtain an edge over their competitors. The recession dragged on into the early 1980s. Nearly all markets, including transportation, were hit hard by a decrease in sales worldwide. Union Switch & Signal's business had one glimmer of hope on the transit side, with new or upgraded systems in Miami, Cleveland, Philadelphia, Baltimore and Atlanta.

In June 1980, Mark H. Sluis was named Vice President and General Manager of the Union Switch & Signal Division, replacing Glenn Stinson. Born in Amsterdam, Netherlands, Sluis (then five years old)

Mark H. Sluis

came to New York in 1931 with his parents. He graduated in 1951 from the University of Michigan with a Bachelor of Science degree in Electrical Engineering and attended the graduate school of business at the University of Connecticut. Sluis served in the U. S. Navy from 1954 to 1956 as a Lieutenant junior grade. He worked at Bendix Aviation Corporation as a Senior Research Engineer, and then joined Pratt & Whitney Company as a Manager of Electrical Engineering. Sluis spent 14 years at General Railway Signal Company where he was Vice President of Engineering before assuming the role of leadership at Union Switch & Signal.

Mark Sluis had a full plate during his time as the head of Union Switch & Signal. On October 29, 1980, the Company signed agreements with the Irish Government to open its operation in Tralee, Ireland – a endeavor begun by Glenn Stinson. The facility manufactured products specifically designed for the Irish State Railways. Thomas C. Linacre, formerly Manager of New Market Development in the Business and Product Planning Department, worked for V. J. Poremba, vice president, Manufacturing, and directed the initial startup for the Irish facility. Linacre also served as the Field Sales Office Representative under D. F. Wert, vice president, International Operations, coordinating the field sales activities between Swissvale and the Irish State Railways.

Back in Pittsburgh, engineers were busy developing new products, including the TAC-100 (Touch Activated Control), the Micro-700 office control system and the MicroCode DC coded track circuit. While some products never achieve success, they are often the predecessors for other lines that do become profitable. Other times, they become an experiment that is better off left on the shelf. Hindsight is sometimes the only

In the early 1980's, Union Switch & Signal made its first attempt at a long-range, DC Coded Track Circuit called "MicroCode." The product was pulled after its third iteration due to an inability to function well with ballast, spurious noise and other "finicky" track circuit conditions.

deciding factor. The TAC-100 was an interface designed to improve data entry for switch and signal supervisory control and for entrance/exit route selection functions. It reduced the chances of human error during manual keyboard manipulations. Target points, graphically depicted on the screen and assigned coordinates, represented function request points. The touch of a finger or stylus on a target point of the Cathode Ray Tube (CRT) broke a light beam and sent the appropriate function signal to the process system. A scan of the coordinates determined the function being requested. Unfortunately, touch-activated screen technology was ahead of its time, and the product was short-lived.

The Micro-700 control system's microcomputer was Union Switch & Signal's first attempt at building a cost- and space-saving computer to control all Centralized Traffic Control (CTC) office functions, such as dispatching, train sheets and code line communications, from a single electronics card file. Using equivalent standard vital relay equipment would have consumed a space ten feet long, seven feet high and two feet deep – 40 times the space of one Micro-700 unit. It was among the first of the company's first microprocessor-controlled products to be installed on several railroads, including the Southern Pacific Railroad and several locations in Canada.

MicroCode was Union Switch & Signal's first attempt at a micro-

The MicroTrax DC- coded track circuits provided rail integrity warning systems for up to five-mile circuits.

processor-based track circuit for long-range, mainline freight railroads. It went through three iterations (MicroCode I, II and III), but it was never fully successful. While the product was eventually pulled, the technology designed laid the groundwork for MicroTrax track circuit, which grew to be very successful.

Ray Franke and Richard Victor were the key engineers in developing the MicroTrax DC-coded track circuit, which extended the length of track circuits upwards of 30,000 feet, far surpassing existing maximum lengths. Longer track circuits meant fewer wayside huts and equipment to buy and maintain. The task of the MicroTrax was to inject a pulsed signal into the rails and watch for the signal to drop out, or shunt, indicating that an approaching train had entered the block. The first successful MicroTrax test installation on the Burlington Northern Santa Fe (BNSF) Railroad detected track shunts thousands of feet away from the interface point, and even reported a broken rail that went undetected by competing test systems.

Union Switch & Signal celebrated its Centennial on June 13, 1981, with a ceremony at the Westinghouse Memorial in Schenley Park. Mark Sluis presented the company's capabilities to dignitaries and community officials at a special luncheon. Antonette "Toni" Ferguson (nee deSabatine), who started working at Union Switch & Signal in 1966, manned one of the tables at the open house. "We had a big open house with people were doing demonstrations," she recalled. "Everything we built, we built here, so we demonstrated a lot of what we were doing. Visitors could go through all the shops and offices. We had the forge shop where we built equipment cases,

Toni (deSabatine) Ferguson

Toni (deSabatine) Ferguson had worked as a professional secretary for about 15 years when she was offered a unique opportunity at Union Switch & Signal. For some, the chance to change careers might have been an incredible opportunity, but the experience for her was the confirmation that she knew where she wanted to be. Ferguson graduated from Robert Morris, which was at the time a community college in downtown Pittsburgh. "It had a strong coursework in accounting and secretarial. I took secretarial courses, earned an AA degree in one year, and started at the Switch right after I finished," she explained. "I failed my employment tests because I was bad at tests. They gave me a chance anyway and came back to me and said I was a 'diamond in the rough.'" There was an era in the 1980s when Union Switch & Signal encouraged women to move out of secretarial positions and into other fields. Andrew (Andy) J. Carey was Vice President of Engineering in 1981. "Andy Carey, my boss, asked me if I would be interested in becoming an engineer," she said. "I didn't want to let him down, so I took the classes for a year. They were training me on the job as an engineer in the hardware/software on the CSX projects. I finally told him that I really didn't want to do it and that I loved secretarial work. I went back to being a secretary, and I've always had excellent bosses. I feel very fortunate."

Guests listen to the presentation at the Centennial Celebration in Schenley Park.

William A. Marquard, Chairman, president and CEO of American Standard, with Mark H. Sluis, Vice President and General Manager at the Westinghouse Memorial Centennial presentation.

and the railroad cars came right into the plant to unload. The relays were built on the fourth floor. The exhibit hall was packed. There were thousands of people (5,093 to be exact) who came through that day. Not just families, but a lot of locals. There were balloons, souvenirs, hot dogs and Cokes. There was even a Conrail locomotive for visitors to see."

The Centennial celebration should have lasted all year, but the company's pride and festivities were overshadowed by an event that would shake the company and change its direction for the years to come. Some of the employees were restless. Shortly before the Local 610 of the United Electrical, Radio and Machinery Workers of America's contract expired on October 31, the union tried to negotiate a new contract for 4,000 workers at the Westinghouse Airbrake plant in Wilmerding and the Union Switch & Signal plant in Swissvale. They demanded for substantial wage increases, a cost-of-living adjustment, supplemental unemployment benefits and protection for workers displaced because of automation. William A. Marquard, Chairman, president and CEO of American Standard, said revenues for 1981 were severely impacted by depressed markets for building products and capital goods around the world and weaker European currencies. In essence, it was not a good time for anyone in the company to ask for more money.

In response, on November 1, 1981, about 4,000 employees walked out when the old contract expired. The union expected a quick resolution, but negotiators were unable to reach a settlement. The union had only had two strikes at Westinghouse Airbrake during the past 40 years, and they had lasted only two weeks.

The striking employees set up picket lines at both plants. The pickets became violent, leading to a rash of broken windshields in employees' driveways; employees threatened with knives; and rocks hurled at cars as they crossed the picket line. Union Switch & Signal reported that non-striking employees had to replace more than 350 vehicle tires and repair more than 800 additional tires as a result of the violence.

On November 5, about 900 non-strikers were allowed to enter the two Westinghouse Air Brake plants, but a judge pronounced in December that Westinghouse Air Brake could only ship equipment that was

About 2,200 United Electrical Workers Local 610 strikers left the East Allegheny High School in North Versailles after voting on American Standard's offer.

produced by union personnel who already had been paid for their work. Non-union workers and supervisors were not allowed to produce anything at the plants. Determined to keep the business going, Mark Sluis rolled up his sleeves and manufactured parts with his non-union employees. His son, Scott Sluis, now an area bank president in Utah, remembered the strike as a young boy. "They threw eggs at my father's company car and yelled at him. My father would come home with remnants of egg splattered on his car from driving past the strike line. As a little child, I remember the eggs."

The strikes affected at least six multimillion-dollar subway and mass-transit projects across the country, including a $1.5 billion subway and surface rail commuter program in New York. Weeks became months and times were difficult for both the strikers and those who crossed the line. Quotes filled with bitterness, frustration and loss splattered across the pages of local and regional newspapers. It began to appear that the impasse might never be resolved. In fact, it was 205 days – one of the longest in the area's history – before the agreement came. At 5:00 a.m. on May, 5, 1982, following three weeks of nearly nonstop negotiations, the union won back benefits lost over the course of the strike and received their raises, albeit they lost an average of $11,000 over the course of the strike. Twenty-three strikers who were fired because of alleged injunction violations would be given the right to appeal their cases through arbitration. But strikers still refused to vote until the company put into writing the arbitration procedure for employees fired. The company agreed to handle the cases through grievance channels after workers returned. The company and union agreed to a time study of some piecework jobs at the Swissvale plant and the company installed a data reporting system to track production and employees.

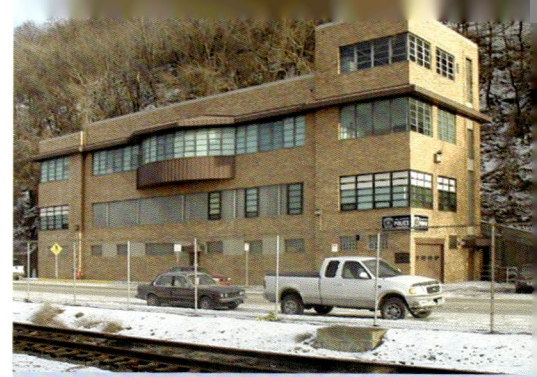

Port Authority.....

The Port Authority of Allegheny County's (PAAC) Pitt Tower at Pennsylvania Station in Pittsburgh was the largest system of its kind – a Union Switch & Signal Model 14 power interlocking machine... with 367 levers. Work began in 1948 but the project was never completely finished as the PAAC removed many tracks over the years, making large portions of the machine unnecessary. The end of PITT Tower came in 1980 when Conrail took out more rail crossovers that in turn required fewer interlockings, and replaced Pitt with several small relay-based control points. Model 14s were first used in 1916. Several 100+ lever Model 14s were installed in the mid-1930s, but only a few were built or modified after 1940. New York City Transit still had about 20 Union Switch & Signal and General Railway Signal power interlocking machines in-service as of early 2009.

Woman assembling printed circuit boards using a semi-automatic, through-hole machine in the Batesburg plant, circa 1985.

Sluis told news reporters after the strike ended that Union Switch & Signal had about 1,700 employees – not far from their 2,000 peak employment level. He emphasized that an economic recovery would go a long way toward restoring full employment, but that competition from smaller firms was eroding the firm's overall profitability. "We're a full-line supplier of highway crossing equipment," he was quoted. "In recent years, we've been faced with stiff competition from smaller outfits that pick out a single piece of the equipment package and manufacture it at a lower cost." Indeed, railroads could now buy individual pieces and assemble their own equipment at costs well below that of purchasing the finished product.

The triumph of the union thus had a bittersweet ending. The company, seeing no possibility of profitability under the new agreement, began to move production to the Batesburg, South Carolina, plant shortly after the strike. Electronics, relays and switch machines were the first to go. By early 1984, a 120-man wiring department, assembly of track switches and signal lights were moved to a new plant in Macon, Georgia. The company used both non-union plants for part of a $19.7 million signal network contract for the Port Authority Transit's (now the Port Authority of Allegheny County) 10.8-mile light-rail transit system in Pittsburgh.

On May 24, 1983, Mark Sluis died suddenly from a heart attack at 55 years old. The company gave his wife Phyllis the company car that he drove – a Cadillac. Sluis had reported to Kenneth Liatsos, who was then Group Vice President Signaling Products of American Standard. Upon Slius' death, Liatsos brought Union Switch & Signal business and its subsidiaries under the umbrella of his group. The same year, major railroads in both Europe and the United States cut back on capital spending as the recession eroded their traffic. Railroad haulage in the United States began to revive late in the year, but it had not yet impacted new orders. Compagnia Italiana Segnali's project for the Italian State Railways continued to support Union Switch & Signal through 1985. When the contract ended and two large mass transit contracts booked in prior years, the Metro Traincontrol System in Seoul, Korea, and the CTC system from Mexico City to Queretaro, Mexico, kept manufacturing going. However, the company finally shut down its forge shop and the steam-generating powerhouse at the Swissvale plant.

Audio frequency (AF) track circuits became the building block for the wayside, office and cab signaling subsystems of Automatic Train Control (ATC), and were modified to suit customer needs throughout the 1980s. The AF-400 was installed on the Baltimore Transit Main Line and the Miami-Dade Metrorail Line in 1983; the AF-400A was used in the Seoul Metro in 1985; and an upgraded AF-400B was sold to the Baltimore Line for their Phase 1, Section B project in 1987. Additional modifications – the AF-500 and AF-600, were installed on the Chicago Transit Authority (CTA) in 1985 and Atlanta's Phase B-2 North-South Line Extension in 1986, respectively.

In 1984, WABCO Westinghouse Compagnia Italiana Segnali in Piossasco (Torino), Italy, was operating at near capacity, but Union Switch & Signal facilities, sustained largely by mass transit systems orders, tapered off. Railroad and mass transit markets remained depressed in 1985. By contrast, in the early 1980s, Union Switch & Signal had provided both light and heavy rail transit control systems for transit authorities across the United States and in the major cities of the world. The company installed Light rail systems in Pittsburgh,

San Diego, Portland, Oregon, and in Calgary, Canada. Heavy rail systems were built for New York, Boston, Atlanta and Chicago, while international systems were delivered in Montreal, Oslo, Bucharest, Stockholm, Sao Paulo, Milan, Dublin, and Seoul.

The AF-500 track circuit made its debut on the Chicago Transit Authority lines in 1985.

In August of 1985, Kenneth Liatsos announced tentative plans to close the Swissvale plant, a local landmark for nearly 100 years, by 1987. American Standard said it would move the Union Switch & Signal corporate headquarters to an office park in Pittsburgh's North Hills. The Swissvale plant was sold in 1985 and the company began to phase out operations, closing the doors for good in 1987. In the spring of 1986, Union Switch & Signal's headquarters was relocated to a five-story office building in McCandless Township, about 20 miles northwest of Swissvale. Meanwhile, a customer service center was established in Augusta, Georgia.

In spite of these moves, the engineers of Union Switch & Signal managed to bring two highly successful products to market in 1985, with the development of the first microprocessor applications. The office-based DDI-601A digital data link and the wayside unit, the Genisys non-vital logic emulator, consumed far less space and power then discrete or relay-based systems. Smaller systems required smaller power sources, and if a circuit board or even the entire unit failed, it was easily swapped out. Genisys made its debut on the Canadian National and Canadian Pacific in 1985. The engineers did not leave the electro-mechanically inclined signal technicians of the railroads to make a cold transition the digital world, either. They provided a user-friendly software development tool that mimicked drawing board type relay logic development on a computer screen to assuage the process.

The same year, Union Switch & Signal field-tested their new MicroLok vital (failsafe) system. The first field test took place at Conrail's Esplen interlocking in Pittsburgh's West End. Norfolk Southern used the new MicroLoks for their Roanoke, Virginia, yard approach interlocking, and within a few years, Amtrak and SEPTA (Southeastern Pennsylvania Transportation Authority) adopted MicroLok based on its reputation as a reliable, failsafe unit. The failsafe element meant that any failure of any test cut the power and downgraded the signaling system to a safe state – the "stop" aspect. The MicroLok significantly reduced the amount of re-quired wayside hardware, wiring and power consumption of earlier relay-based systems. Despite MicroLok's wave of popularity, it began to experience unexplained shutdowns. It took several years of field-testing for technicians to identify and solve the anomalies. Union Switch & Signal engineer Ronald Capan, one of the signaling industry's leading experts on transient voltage protection, developed several practical solutions to the problem. The microprocessor-based controllers were hypersensitive to even the smallest stray

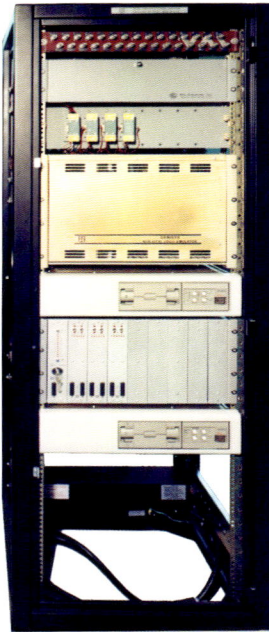

The non-vital Genisys provides office-to-field coded track communications.

Union Switch & Signal collaborated with Conrail in 1985 to field-test the MicroLok system at a low-traffic site in Pittsburgh's West End, the Esplen interlocking.

electrical currents – from lightning strikes miles away to the wayside engineers' chatter on walkie-talkie radios.

The demand for Union Switch & Signal's products and systems in the United States, especially railroad CTC systems, was strong in 1986. The company delivered the first CTC office (Union Pacific's Harriman Dispatch Center in Omaha) to combine automation of train functions and dispatcher reports with video display of traffic and signal systems for the railroad's entire territory. CSX soon ordered a similar system for its Dufford Control Center in Jacksonville, Florida, the largest office installation at that time. Six other railroads began to replace their relay-based interlockings with MicroLok, and Conrail integrated MicroLok with other electronic control devices.

In October of 1986, the Italian State Railways awarded American Standard the largest single signaling order ever placed by the agency, to the tune of over $80 million for a three-year contract. Union Switch & Signal's Westinghouse Air Brake Westinghouse Compagnia Italiana Segnali opened a new plant later that year in Piossasco (Torino), Italy, to serve the Italian State Railways' long-range modernization program.

An Italian railroad supply company called Ansaldo Trasporti S.p.A. acquired Transcontrol (established by the former Union Switch & Signal exec-

Ronald Capan

Trains await to be classified at Roanoke Yard, Norfolk Southern Railroad in Roanoke, Virginia.

"…Ansaldo Trasporti had agreed to acquire the railway signaling operations of American Standard Inc." ….. Ansaldo would acquire the Union Switch & Signal Division in the United States along with American Standard's Italian-based WABCO Westinghouse operations.

utive, John Pelikan) and several of its employees, including Edward Riddett and Kevin Riddett, in 1987. Ansaldo Trasporti was a wholly owned subsidiary of the Italian mega-conglomerate, Finmeccanica Societa Finanziaria Meccanica S.p.A. Headquartered in Islip, New York, Transcontrol was a leading supplier of systems engineering services and signaling, switching and control products – primarily for rail-transit applications.

On July 4, 1988, Finmeccanica announced, "…Ansaldo Trasporti had agreed to acquire the railway signaling operations of American Standard Inc." Finmeccanica, a unit of the State holding company Istituto per la Ricostruczione Industriale, said that Ansaldo would acquire the Union Switch & Signal Division in the United States along with American Standard's Italian-based Westinghouse Air Brake Westinghouse operations. The purchase price was $105 million. The same year, American Standard sold the Westinghouse Air Brake Division to SAB (Sweden). SAB later sold the North American Westinghouse Air Brake operations to management in 1990.

Prior to the acquisition, Arthur "Art" W. Ticknor, Vice President and Group Executive Fluid Power Products for American Standard, succeeded Liatsos in August of 1987 with a new, combined title of Group Vice President of the Signaling & Fluid Power Group of American Standard, which included Union Switch & Signal and its Italian signaling company subsidiary, Westinghouse Air Brake Westinghouse Compagnia Italiana Segnali. In 1987, Ticknor moved the Fluid Power Group from Lexington,

Ansaldo S.p.A "sealed the deal" to purchase Union Switch & Signal July 29, 1988. (left to right) – Alberto Rosania, senior advisor of Finmeccanica; Nicholas N. Georgitsis, senior vice president of American Standard; Preston Tollinger, attorney for Coudert Brothers.

Arthur "Art" W. Ticknor

Kentucky, to Pittsburgh. He joined American Standard in 1962. From 1968 to 1971 he worked as a Materials Manager at Union Switch & Signal and then worked at Westinghouse Air Brake in Gateway Center. In 1972, he moved to Buffalo, New York, after American Standard acquired Westinghouse Air Brake. After the acquisition by Ansaldo on July 29, 1988, Ticknor relinquished all of his other Group Executive duties and was appointed President and CEO of Union Switch & Signal.

Senior Advisor Ing. Alberto Rosania of Finmeccanica S.p.A. was the key senior executive who, in 1988, strategized the acquisition of Union Switch & Signal. He vividly remembers the transaction. "I was Executive Vice President of Ansaldo Trasporti, and realized our signaling [business] was fairly weak. Ansaldo Trasporti made electrical parts of the rolling stock, power supply and signaling, but we didn't have advanced technology in signaling at the time. Union Switch & Signal was interesting for several reasons – their technology was advanced. Germany and France also had the technologies, but they had nothing for us to acquire. We acquired a minority share (40 percent) of a small, but brilliant company called Transcontrol in 1987. The market was dominated by Union Switch & Signal, GRS, Safetran and Harmon. GRS and Union Switch & Signal held about 70 percent [of the market]. We got news that American Standard was going to dismiss its interest in the signaling and brakes. In 1987, I contacted American Standard and learned that they were interested. We started to talk to them but nothing was done until 1988. In 1988, they decided to sell. We gained 30 percent of the American market with Union Switch & Signal. The interesting thing for us in buying Union Switch & Signal was that we got hold of their Italian subsidiary, Westinghouse Air Brake Westinghouse Compagnia Italiana Segnali and their brand new plant in Piossasco (Torino), in southern Italy. They had 30 percent of the Italian market. We were in Naples and Genoa with 30 percent of the Italian market, so then we had over 60 percent of the Italian market. We became the most important [rail supply] company in Italy."

"American Standard couldn't see how successful the business could be," Art Ticknor postured, "and they sold it. American Standard sent [Union Switch & Signal] away in 1988, but went private and had to find cash, so they decided to spin it off and they called Ansaldo. It was a quick acquisition. [Glenn] Stinson tried to buy it, as did several other entrepreneurs." But Ansaldo gave American Standard the best offer.

When Art Ticknor was appointed President and CEO of Union Switch & Signal, the company had been through the turmoil that comes after a lot of high turnover, as well as some big problems with changing wayside equipment from electro-mechanical to solid-state technology. Ticknor recalled, "The first MicroCode units on the Boston and Main and the New York Central were failing due to lightning strikes, so we had manufacturing issues. Harold Gillen was one of the engineers who helped to fix the issues.

Ticknor said that there was no General Manager position at that time. "I was President and effectively operated as the General Manager, but without the title." The biggest projects during Ticknor's leadership were the Burlington Northern and the CSX control centers.

The Union Pacific Harriman Dispatch Center in Omaha, Nebraska, continues to supervise over 32,000 miles of territory, using 151 code lines, 9,600 non-CTC control locations, dark territory control and Rule 251 territory control. The CSX Dufford Control Center in Jacksonville, Florida, also continues to control 17,500 miles of territory with 37 dispatchers. The Jacksonville center manages 177 code lines, 7,200 TCS (2,300 locations over 7,400 signal pairs), 7,700 DTC manual block locations (dark territory), and 2,600 train order and

The Union Pacific Harriman Dispatch Center in Omaha, Nebraska, supervises 32,000 miles of territory using a Union Switch & Signal office system.

miscellaneous locations. Today, Union Switch & Signal's CTC technology clears in excess of 160,000 signals a week for CSX alone! "The CSX facility is a fascinating place that looks like a silo," Art Ticknor said. "Inside, it is dark with computer screens in the center where the dispatchers sit and around the periphery, are the track layouts throughout the system. When I first went there, we had an appreciation that components had higher margins but systems drove the sales of those components and repair parts down the road."

Crawford "Tacky" Staples started in 1931 as a recent graduate railway electrical engineer making $135 a month. Staples wrote his thesis on the history and development of railway signaling while studying at the University of Illinois at Urbana and contacted Union Switch & Signal's Chicago office as part of his research. This connection led to an apprenticeship at Swissvale. He was assigned to the CTC area, then transferred to train control systems where he worked on the first major coded track installation. He also worked for several years in the train control lab on equipment design.

Staples spent the next 20 years as a field engineer. "I got involved in systems design, mostly coded track, during an extensive period of development," he said in a 1980's SwitchPoint newsletter interview. Of the 64 patents credited to his name, 40 were put into service. Tacky Staple's most valuable contributions included the phase selective coded track circuit, the ferro-resonant track circuit, and normally de-energized and slow coded systems. Staples became Supervisor of Coded Track Systems, then in 1953 was named Section Manager of Track and Line Circuit – a department which was expanded and re-named the System Analysis Section, and later emerged as the Electrical Section. In the early 1960s, he became a Consulting Engineer in the Research Department and led the task force for the first digital computer yard in Eugene, Oregon. He was always available to share his expertise with other Union Switch & Signal engineers. He is still considered one of the world's leading track circuit experts.

A Union Pacific Passenger Special races west towards the Bay Area as it takes passengers from the small mountain town of Portola, California to the busy streets of Oakland, California.

Photo Credit: © Jake Miille

14   The Drive for Driverless

More legendary railroads disappeared into mergers between 1990 and 2000, most notably the Atchison, Topeka & Santa Fe Railway. In 1994, the Santa Fe and the Burlington Northern formed the Burlington Northern Santa Fe Railway, now known as BNSF Railway. In 1995, the Chicago & North Western Railway became part of the Union Pacific. The merger of the Union Pacific and Southern Pacific railroads in 1996 yielded the largest railroad in North America, a network of 32,000 miles of track across 23 states. The Canadian National purchased the Illinois Central Railroad in 1998, and in 1999 Conrail was split up between CSX and Norfolk Southern, leaving North America with seven major Class 1s: Burlington Northern Santa Fe Railway, CSX, Kansas City Southern, Norfolk Southern, Union Pacific, Canadian National and Canadian Pacific.

Walter Alessandrini

Business for the railroads showed improvement, and the economy in the United States was finally breathing again. Union Switch & Signal landed a myriad of significant contracts and developed industry-first technologies and products that soon repositioned the company at the forefront of the transportation industry. In spite of the positive market conditions, frequent changes in management and management philosophies created a very instable atmosphere. President and CEO Art Ticknor was appointed Chairman of the Board in 1990, succeeding Edward Riddett. Walter Alessandrini, who had served as the Executive Vice President and Chief Operating Officer, became President and CEO of Union Switch & Signal. Alessandrini was born in Chiavari, a town on the Italian Riviera, and graduated from the nearby Universita' di Genova (University of Genoa) with a doctorate in Mechanical Engineering. After serving the mandatory conscription military service, he landed a job with Varian, a Palo Alto, California based semi-conductor company with a division in Italy. Following his initial employment with Varian in Italy, he transferred to the California headquarters where he remained for a period of time. While in California, he learned that Ansaldo STS was looking to expand their transportation business into the United States. In the July 1995 issue of the Greater Columbia Business Monthly magazine, he was quoted, "they did not have any experience in the United States and I didn't have any experience in the transportation business, so I guess it was the perfect match."

Alessandrini started working for Union Switch & Signal after the Ansaldo acquisition. His new staff included Jim Ausefski, Vice President of Engineering and Kevin Riddett as Vice President of Manufacturing.

Who is Ansaldo?

In 1854, Ansaldo Genoa-Sampierdarena Works designed and built Italy's first steam locomotive. Today, the company is involved in Italy's high-speed railway system and is the licensed operator and supplier of subsystems for most of the 33 regional Italian railways. Ansaldo STS designs and builds electric railways and urban transit systems worldwide and is a world leader in the production of signaling and automation systems.

Ansaldo is a subsidiary of Finmeccanica, which was owned by IRI, a holding company. Its subsidiaries, in addition to Union Switch & Signal, include Transystem and Segnalamentao Ferroviario in Italy, AT Signalling in Ireland and AT Signal System in Sweden. Ansaldo Trasporti at the time, along with Ansaldo Industria and Ansaldo Energia, were all part of Ansaldo, a company with nearly 150 years of experience in railroad technology.

Giovanni Ansaldo

Kevin Riddett was sent to manage the Batesburg operations, replacing Anand Sharma. Riddett expanded the facility in Batesburg and soon closed the operations in Augusta, Georgia transferring all the personnel to the Batesburg facility.

Four Union Switch & Signal engineers, Dave Cichowski, Bob Brey, Paul Herre and Brad Guest recognized the advantages of the MicroLok and the Genisys and adapted the technologies to the first-ever vital, microprocessor-based cab signaling system, which was named MicroCab®. The space saving MicroCab replaced bulkier equipment racks in the confined spaces of the locomotive cab. The Union Pacific Railroad ordered 72 vital MicroCab sets in 1991, and soon after, mass transit agencies across the country were placing orders for hundreds of units, including the Massachusetts Bay Transportation Authority (MBTA), New Jersey Transit (NJT), Long Island Railroad (LIRR), Metro-North Commuter Railroad (MNCR) and Chicago Transit Authority (CTA). In 1998, 39 sets were installed on Amtrak's Acela high-speed trains on the Northeast Corridor. By the late 1990s, MicroCab was established as a solid platform for vital Automatic Train Protection (ATP). The MicroCab was augmented to serve as an onboard event recorder, send its stored data to a separate event recorder, communicate over a Local Onboard Network, perform contingency braking actions and detect when train wheels were slipping or sliding.

The Union Pacific Railroad was the first to install the new MicroCab, a first-to-market vital, microprocessor-based cab signaling system, in 1991.

In 1991, the Los Angeles County Metropolitan Transportation Authority (LACMTA) awarded Union Switch & Signal a contract to design build and install the first driverless light rail system in the United States. This Automatic Train Control (ATC)

system integrated MicroLok, MicroCab and the AF-900 track circuit for the first time into an all-vital microprocessor way-side/cab signal Automatic Train Protection (ATP) subsystem. In a 1992 KDKA television interview with financial section Bill Flanagan, Alessandrini talked about the new Green Line's driverless system. "It is the very first driverless system in the United States, and a showcase of technology. The whole world is actually looking at this new piece of technology going into Los Angeles. Driverless might mean more safety. If you look at the past history of accidents that sometimes we experienced in some of our urban transit systems, in most of the cases they were due to the human error, so the driverless system is actually safer. A big part of it is based on technology that has evolved over the past years that is proven for reliability and safety. The management will be controlled by one central control center." The system was to be installed on LACMTA's Green Line, but it never evolved into a fully driverless operation

A Los Angeles Metro Green Line Train in Redondo Beach, California, runs safely with a Union Switch & Signal onboard cab signaling system, installed in 1998.

due to several project and financial hurdles. It would, however, serve as cornerstone for ATC projects in the next decade that did achieve full driverless operation, most notably the Copenhagen driverless system.

More projects rolled in from the international side, with an expanded contract on the national Railway of Mexico's Queretaro Electrification Project, and the Orbassano Classification Yard project in Orbassano, outside of Turin, Italy. Called a "marshalling yard" in Europe, it consists of 40 bowl tracks and 50 miles of rail for sorting and assembling trains. Stateside, Union Switch & Signal also began work on the Washington Metropolitan Area Transit Authority (WMATA) "E" line – a project that controlled train movement on 7.95 miles from Fort Totten to Green Belt Station. The project included 17 switch layouts, 22 signals, 154 impedance bonds, 2,083 relays and an array of AF-800 track circuits.

The year of 1992 was a soaring success for Union Switch & Signal. They won a transit control contract from the Seoul Metropolitan Government for three of its subway lines, upgrading and extending the Central-ized computer control office operation, and an installation of 92 MicroCab signaling systems. The first MicroCab was installed on a mass transit system on the MBTA. They were awarded a contract from the Italian State Railways to supply equipment and assist in the design and engineering of the new Cervignano classification yard, and they installed the first MicroTrax for revenue-service at Hammersley Iron, near Perth, in Western Australia.

The shipping warehouse in Batesburg comprises 67,830 square feet.

Right: Product Specialists Tim Spangler, left, and Tommy Adams (now retired), right, assemble an M-23 switch machine in the Batesburg Ground Equipment area of the plant.

The Los Angeles County Metro Transit Authority Green Line was the first to install the AF-900 coded track circuit in 1994 on their light rail transit line.

"In 1990, Bob Kull, working on behalf of the Hamersley Mine in Northwest Australia," said Ray Franke, Union Switch & Signal Section Manager, "wanted to know how long we could stretch MicroTrax. I proposed a low-frequency cab signal system with 40-cycle cable signal overlaid on MicroTrax over 40,000 feet. Our Swedish company did the onboard and we did the wayside. It was installed 15 years ago, runs 43,000 feet, and still runs like a champ. That branched into the Copenhagen job, Thessalonica, China, and others, all based on the AF-900. It was a terrific product."

Union Switch & Signal acquired the product lines and business assets of Dynamic Sciences Limited (DSL) of Canada in 1992, the industry leader in end-of-train telemetry systems and an innovator in locomotive simulation technology. That summer, the company relocated the Union Switch & Signal corporate headquarters to Columbia, South Carolina, to "complete an internal reorganization to focus on three key business areas: components, systems, and service and training." Pennsylvania retained the system engineering and R&D with about 450 employees, and the manufacturing remained in Batesburg. Part of the concept of the relocation to Columbia was the support that Union Switch & Signal would get from the University of South Carolina's engineering department in fostering new signaling technologies.

The year continued to be a success when in September 1992, Union Switch & Signal was the first full-line signaling products and systems supplier in the U.S. railway industry to receive ISO-9001 Registration. This coveted international business community's recognition of quality processes provided the company with the international "stamp of approval" in regards to its quality system standards. It was a red-letter day, to be sure.

Walter Alessandrini had been with Union Switch & Signal for five years, three of which he was President and CEO. An excerpt from a mid-1993 editorial to the employees acknowledged the company's successes. "Little did I know that by July of 1988, I would be in the middle of Union Switch, charged with turning the company inside out as the acquisition unfolded. Through our talks I realized that turning the Switch around would require teamwork, good leadership, focus, empowerment and hard work – no magic wands, nothing revolutionary. The meetings with our customers were painful. Slowly but surely, the success came. The results are impressive: back to profitability in two years. A 50 percent increase in productivity. Double the amount of sales in four years."

Indeed, the company was flourishing. The new Genisys 2000 was ready for market. The first digital AF track circuit, the AF-900, was installed on the LACMTA Green Line. Mass transit jobs were on the rise with new system installs from the Seoul Metro lines in Korea to the Dallas Area Rapid Transit (DART) Light Rail Starter System in Texas.

Burlington Northern Railroad awarded Union Switch & Signal a state-of-the-art railroad dispatching control system project in Fort Worth, Texas. The system was to be part of the most advanced dispatch facility in the United States rail industry, using Union Switch & Signal's advanced Computer-Aided Dispatching (CAD) software in its new Network Control Center. The system would automatically route trains based on schedule and priority level, with dispatchers only needing to specify the origin and destination of shipments. Union Switch & Signal's project manager J. T. Miller ensured that the railroad's more than 400 dispatchers received simulation-based training to familiarize them with the functional operations of the new system.

The emerging Chinese market also brought multiple contracts for Union Switch & Signal. The company secured contracts to provide control systems at the Xuzhou classification yards, located about 500 miles north of Beijing. Two years later, contracts were signed for systems in the cities of Fuyang and Xiangtang for a new railway from Beijing to Hong Kong. And in 1997, the company won the Automatic Train Control project on the Shanghai Metro.

Union Switch & Signal broke ground on their $20 million Systems and Research facility in Pittsburgh on September 23, 1993, the same day they announced the initial public offering of Union Switch & Signal stock. President and CEO Walter Alessandrini was quoted in the corporate newsletter, "A company which five years ago was on the brink of bankruptcy is now stronger than ever. By being able to go public, we as a

The emerging Chinese market includes the 1997 Automatic Train Control project on the Shanghai Metro.

Union Switch & Signal President and CEO Walter Alessandrini boards a bulldozer at the ground breaking of the new Union Switch & Signal Systems and Research facility at 1000 Technology Drive in Pittsburgh. Looking on from left to right: Mike Pracht, vice president, Systems Operations of Union Switch & Signal; Herb Packer of the Governor's Emergency Response Team; Sophie Masloff, Mayor of the City of Pittsburgh; Robert Behrabian, president of Carnegie Mellon University; and Dave Jannetta, secretary of the Department of General Services.

company will have access to the capital we need for our growth." The public offering of 3.25 million shares of common stock at an initial $15 a share represented 40 percent of the shares. Ansaldo Trasporti S.p.A. of Italy held the other 60 percent.

Union Switch & Signal submitted a proposal for a new technology to a competition sponsored by the United States Department of Commerce's National Institute of Standards and Technology (NIST) in 1994. The competition, established in 1990, awarded grant money to businesses for the research and development of promising but high risk technologies that could greatly enhance economic growth in the United States. After two elimination rounds, the Advanced Technology Group (ATG), a research and development department at Union Switch & Signal, won a $2 million grant in 1995. Frank Boyle, who was technical manager of ATG, led the project in developing a technology called Asynchronous Teams, or A-Teams™, architecture, which was a computer software framework designed to optimize scheduling and on-time performance in the transportation industry. The company contributed a cost share of $967,000 to the project.

The architect's conception of the new Union Switch & Signal facility.

Union Switch & Signal continued its expansion into the upcoming Australian market. In 1995 the company acquired a company named Ventura. Ventura was a small upcoming transportation company founded in 1983 by Lyle Jackson. Ventura represented and distributed Union Switch & Signal products on the Australian market. The deal was a perfect match for the strategic growth of Union Switch & Signal. Lyle Jackson was named Managing Director of the newly formed Union Switch & Signal Asia Pacific operations. The defined territory of this new operating unit included Australia, India, South East Asia, and Central and Southern Africa. The engineers at Union Switch & Signal never sat on their laurels for long. A team led by Dan Disk, Ray Franke, Judi Kozlowski, Rich Victor, Fred O'Leary and Rich Vehar, reengineered the original MicroLok into a new product that would in time be recognized as the industry's leading microprocessor-based wayside controller. They incorporated many of non-vital office elements of the Genisys product and local control panels. MicroLok II absorbed the best features of MicroTrax and eliminated the need for a separate non-vital Genisys unit for office to field code communications. The controller board is the same on a Genisys II and on a MicroLok II, so the most common non-vital input/output (I/O) boards are all interchangeable, but with different operating systems. The first MicroLok II went into revenue-service on the MBTA Green Line Rehab project in 1996.

Walter Alessandrini left Union Switch & Signal in December 1996. What followed over the next three years was a hyper-drive succession of leadership at Union Switch & Signal. With those uncertainties in management came a period of financial difficulty for the company, a difficulty that protracted well into the next decade. Kevin E. Riddett held the position of President and CEO for four months until March of 1997,

The MicroLok II absorbed the MicroTrax and Genisys features to provide an all-in-one package.

followed by James N. Sanders, a German who had been working for Ansaldo in England and Italy. He held the position for about seven months.

On September 1, 1997, Gary E. Ryker, erstwhile executive vice president of Marketing, Sales and Services of competitor Harmon Industries, was recruited and appointed the new President and CEO of Union Switch & Signal. Ryker had worked in the industry starting in 1983 at Rockwell Collins, a company that was installing the Advanced Railroad Electronic System (ARES) with Burlington Northern (before they merged with Santa Fe). "That development led to the first train control system using GPS for train location," Ryker said. "The BN was trying to be forward thinking. They hired Stephen Ditmeyer as the Chief Engineer of Research & Development. His charter was to find new technologies to solve railroads problems. The ATCS [Advanced Train Control System] process started around 1982, reaching its height in the mid 1990s. Most of the traditional signaling suppliers started delving into this type of system in varying forms in the mid 1990s, including Harmon, GRS and US&S. We demonstrated the product in 1984 and 1985; it later evolved into the industry's ATCS effort, which in turn evolved into ITCS at Harmon."

A technological and product evolution related to the enhancement of the capabilities of the computer-aided dispatching systems occurred at Union Switch & Signal during Ryker's tenure. "There were some preliminary efforts of including dynamic scheduling [that] allowed the dispatcher to react to the demands of the day and any unusual occurrences in a more timely fashion," Ryker explained. "If you had a rail buckle in the southwest in the summer, they wanted to help the dispatcher reroute trains. Several manufacturers were trying to do it, primarily for UP and CSX. There was also a big transition going on from relay-based signaling systems and microprocessor-based signaling systems. They were evolving the MicroLok and competing with Harmon's VHLC business – also a microprocessor interlocking cabinet. General Railway Signal also got into the fray, but not as quickly. That lasted for several years, transitioning from hardwire systems to microprocessor-based to use Boolean logic. The technology…was expensive and a radical departure. If you're responsible for safety, you want to ensure that what you provide is safe and does what it's supposed to do. It's a lot of leap to new technology. If it's wrong, it falls on your shoulders. If you take a technological leap in the railroad safety industry, someone may die. That slows down the technology progress, as it should."

Gary Ryker left Union Switch & Signal in June 1998, about one year after his appointment. Ansaldo S.p.A asked James Sanders to return and fill the vacancy for an interim period. During this time, the Ansaldo S.p.A Board of Directors made the decision to repurchase all of the privately held of Union Switch & Signal stock. By the end of 1998, the stock acquisition was complete and the company reverted to a fully owned company of the Finmeccanica Group. John Mandelli was appointed in August of 1998 as President and COO of Union Switch & Signal and remained in office until July of 2000. The Union Switch and Signal, Ansaldo S.p.A. and Ansaldo Breda companies now constituted the Transportation portfolio of the Finmeccanica Group.

As safe and efficient as track-circuit based train control systems had become by the 1990's, the global mass transit industry pursued an entirely new approach under the banner of "Communications-Based Train Control," or CBTC. This revolutionary concept in train control uses radio systems to exchange vital and non-vital data between the wayside and the moving trains. CBTC had its fair share of growing pains through the 1990's, forcing many transit authorities to keep traditional track-circuit based automatic train control as an underlay or fallback to ensure reliable and continuous service. However, toward the end of the 2000's, many of the initial anomalies were worked out. Ansaldo STS capitalized on these improvements by addressing the fast-emerging market in China. With its technology partner Insigma, based in Hangzhou, China, Ansaldo STS secured major programs for the municipal authorities of Shenyang, Chengdu, Xian and Hangzhou.

Women of the Switch

In a male-dominated industry, women found their niche by proving themselves as skilled professionals who quickly gained the respect of their male colleagues. The stories are wide and varied, but the following women give insight to how they succeeded at Union Switch & Signal.

Mary Eva Spence
Relay Assembler to HR to Reception

Mary Eva Spence holds the record as the longest-serving employee at Union Switch & Signal /Ansaldo STS USA – 47 years as of July 2010 – the longest period of employment when "Safety on the Rails: The Union Switch & Signal Story" was published. She came in during an exciting new era of technology – building micro-miniature relays for the NASA space program. She recalls the "white room" experience – working in "long white smocks and covering our heads." When the plant was constructed in Bates-burg, candidates had to travel to Columbia, South Carolina, to take a skills test. If they passed, the company offered a two-week unpaid training course with no promise of a job. "If you worked out, then you got a job when the plant came in," Spence said. "The only thing we had was the Burlington cotton mill, so the plant was an asset to the community.

"The day I came to work here was the most fortunate day of my life."

– Mary Eva Spence

We made two-way radios and relays. It was mostly women who were employed in the relay department. I worked in the relays, sealing the 900-series of micro-miniature relays." Spence worked for Paul Pollack, who had transferred from Pittsburgh.

"They asked me to move into Accounting as a Clerk in 1965. I was made Personnel Clerk in the Human Resources Department in 1973, where I signed up the new people with insurance and other benefits, and took applications," she said. Mary Eva Spence stayed in the Human Resources department as an Administrator until 2008, when she became the Receptionist.

"One plant manager, Mr. Robinson, said many years after he left, that he was sent down from Pittsburgh to close the plant, but he was determined to make it work. He worked with the plant manager, George Moore, and got it back on its feet."

Judy Kozlowski
Software Engineer

Judy Kozlowski, software engineer has been a central player on all modern-day wayside and cab systems – some of her colleagues call her "a software genius." She took the company from relay logic to software-based logic during the during the transition. Ko-zlowski's accomplishments as an intern won her a job in 1984 as a software engineer in the Swissvale plant before moving to the North Hills in 1986, then into the current building near downtown Pittsburgh. "I enjoyed what I was doing – developing new prod-ucts," – developing new products," said Kozlowski. "PCs weren't that big. I wasn't interested in working on a computer – I liked working on a product that was actually going to do something. We have to do the manipulation and know how the hardware works, even though we don't work on it."

Judy Kozlowski's first project was the original non-vital Genisys, which is still in service in many places today. She also contributed to the MicroLok controller and the original AF-900 track circuit in the early 1990s, and then helped to design the software for the MicroLok II and the Genisys II systems in 1996. In 27 years, MicroLok has been one of the biggest challenges of her career. "There wasn't a micro-processor in the railroad industry for vital interlocking control at the time. So we were breaking new ground. We had to determine what we wanted to do, how it was going to be done, especially from the safety critical aspects. The Genisys taught us a lot, but making it failsafe took a lot of work. You spend a lot of time saying 'what if'. You can make non-vital function with no bugs in it and it works perfectly, as long as nothing goes 'wrong' so it isn't fail safe."

Another challenge was helping the maintainers to accept the transition from relays to electronic equipment. "The mechanical relays might have been a pain to work with at times," Kozlowski said, "but they knew what they were doing because they had worked on them for years. It was 'the devil you know' with the relays. Suddenly, they got a box with flashy lights installed. These railroad men didn't grow up in the computer generation, nor they didn't want to change over to the new products in the first place, and there's this little girl with a ponytail explaining that 'this really will work!'"

Stacey Fedorka
Senior Vice President and CIO

When Stacey Fedorka was 10 years old, her father moved the family from Brooklyn, New York, to Pittsburgh, Pennsylvania, while working for Westinghouse Electric. Shortly thereafter, they packed up again and relocated to Rio de Janeiro, Brazil. It was in Rio where she learned Portuguese and Spanish while in high school. "That really shaped me to go to college to be a political scientist and study international affairs," Fedorka recalled. "I got my Bachelor of Arts in Political Science from Indiana University of Pennsylvania and my Master of Public and International Affairs: Public Administration and International Affairs at the University of Pittsburgh, with the intention of working for the State Department's Foreign Service exam. I thought I would become an ambassador, but two of my professors in grad school told me if I wanted to work for the government, I'd have to compromise my morals. So I decided I didn't want to do that."

Fedorka was overqualified for many jobs, and lacked experience for others. She worked at Aluminum Company of America (Alcoa) in Pittsburgh as a temp and they asked her if she wanted to learn about computers. "It was challenging back then," she recalled. "Computers were not as intuitive. I learned to solder boards, build and rebuild computers, run networks and work on process control boards. I never took a class in computers. I had to learn all on my own."

> "I will be forever grateful for all of the professional opportunities afforded me by Union Switch & Signal."

While she did well at Alcoa, she decided she should gain more business experience. A friend at Union Switch & Signal told her they were hiring. "I took a risk and started in 1998 as a senior technical writer when John Mandelli was president. I was assigned to projects and delivery. I had never done that before. My career continued to evolve." Ken Burke became president and asked Fedorka to consider taking over IT. They wanted to implement a SAP for and ERP (Enterprise Resource Planning) system. "Once I took over IT, I realized the business experience made me a much better IT person. In 2008, I won the Pittsburgh CIO of the year." Stacey Fedorka became the first woman CIO of Union Switch & Signal and, in 2008, the company's first woman C-level officer.

Ron Shaw, Union Switch & Signal sytems engineer studies the action at the Boston CTC Control Center for the Massachusetts Bay Transportation Authority in 1997.

15   Branching Out and Moving Forward

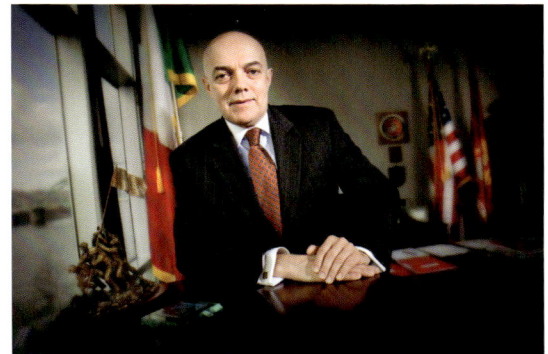

Dr. Alan E. Calegari

The rail industry was doing well in the early to mid-2000s. Freight car sales were up, as were investments in signal communication, waysides and maintenance. The United States government implemented short line tax credits and Railroad Rehabilitation and Improvement Financing (RRIF) loans to give railroads a financial boost for infrastructure improvements. According to the Federal Railroad Administration (FRA), RRIF loans were available to anyone who needed capital, including railroads, state and local governments, and corporations. The FRA permitted them to "acquire, improve or rehabilitate intermodal or rail equipment or facilities," including rail, bridges, yards and shops; the refinancing of debt for the aforementioned projects; or to establish new intermodal or railroad facilities.

In mid-2000, Finmeccanica, the Italian holding company that owns Ansaldo Trasporti and ultimately Union Switch & Signal, was privatized and no longer owned by the Italian state holding company IRI. IRI was subsequently disbanded to comply with European Commission requests.

While Union Switch & Signal was successful from a product development and project standpoint, the issue of an unstable leadership that affected the company for years continued. John Mandelli left the Presidency of the company after two years. His successor Giovanni Roberto Gagliardi, President and CEO of the Italian holding company Ansaldo S.p.A., stepped into the position of President and CEO of Union Switch & Signal in June 2000. In the same time Derek Hume assumed the COO position from June through December 2000. A search for a President and CEO was finalized in February 2001 with the nomination of Kenneth "Ken" R. Burk into the combined offices of President, CEO and COO. Burk brought in John McGee to serve as COO. McGee left the company in 2008 several months after Burk ended his tenure in June 2007.

Ansaldo S.p.A., the holding company of Union Switch & Signal elected Sergio De Luca as Group President and CEO in Italy over all the Ansaldo companies. De Luca immediately understood the criticality he was facing with the underperformance of Union Switch & Signal and initiated a search to fill the vacancy of the President & CEO position in the U.S. subsidiary. During this time, he appointed Emmanuel Viollet, a vice president of Sales for the French subsidiary of Ansaldo S.p.A. as the interim Union Switch & Signal President.

Dr. Alan E. Calegari was appointed President and CEO of Union Switch & Signal in July 2008 after completing his Presidency of Dedicated Micros, Inc., a technology leader in video security applications. Prior to his appointment, Calegari made an in-depth study of Union Switch & Signal and prepared a plan for its turnaround. In a meeting with the leadership of Ansaldo S.p.A., he presented his findings that "the company's issues are only related to the unstable leadership and lack of focus on customers. The company has a strong technological know-how and is well poised for growth in the domestic and international markets."

Alan Calegari initiated a thorough examination of the company's activities and prioritized restructuring actions on some of the more critical programs initiated in earlier years. He soon realized the task of a complete organization turnaround required a team who could examine the details of neglected or misguided operations and redirect the company to achieve a positive financial outcome. "It was reaching midnight on the evening of September 9, 2008, and I was reflecting on the formation of the best unit that could serve with me to reach the high grounds quickly. (Calegari often referred to such maneuvers based on his prior Marine Corps background.) We had good professionals and very capable individuals who only needed to be repositioned in critical functions, be motivated and be empowered to execute. I needed individuals that could innovate every step of the way and that understood the principle of aiming at a 70 percent solution, rather than waiting for a late perfect plan."

M-23E Switch Machine on the Charlotte Area Transit System (CATS) has been in revenue-service since 2007.
Photo credit: Joanne L. Harris

The M-23E machine features an electronic circuit controller (ECC).

Calegari selected the team through the principle of "hire via trial by fire." Time was of the essence and a team that could bond and overcome obstacles needed to be put in place quickly. The night of September 9, 2008 was the night. The team of senior executives that became known as the SES (Senior Executive Staff) included John G. "Jack" Borman as Vice President of Finance and CFO; Stacey Fedorka, Vice President of Technology and CIO; Jeremy Hill as Vice President and COO; Thomas P. Lawton as Vice President Legal and General Counsel; Joseph D. "Rocky" Eiseman as Vice President Manufacturing in Batesburg, and Jeffrey Wang, Vice President and General Manager China Operations. Jeremy Hill left the company the following year.

"Together we worked diligently to mount a clear communication to employees defining their roles and responsibilities," said Calegari. "The results were astounding. Everyone pulled together and we were on the way of a successful year-end closing on the first year." With the team in place, the turnaround plan underwent a quick execution. At the closing of 2009, the company posted record profitability and the largest backlog of projects since its founding. Legacy issues were under control and the race for an even greater 2010 was at hand. The tenure of Dr. Alan Calegari saw a "renaissance" within the company. Union Switch & Signal, under his leadership, achieved a complete financial turn-around in the first 12 months, positioning the subsidiary as one of the best performing companies of the Ansaldo Group and paving the way for the company to enter into the global organization plan.

Jeffrey Wang (left) and Dr. Alan Calegari (right) at the opening of the Chengdu Metro, in 2010.

In May 2002, Union Switch & Signal was awarded the contract to provide signaling and control systems for the Port Authority Trans-Hudson (PATH) in Jersey City, New Jersey. PATH rail lines link Manhattan with Newark, Hoboken and other New Jersey locations just west of the Hudson River. The systems were part of three projects underway to restore or repair the facilities and track that was damaged or destroyed during the September 11, 2001, World Trade Center attacks. The systems included the MicroLok II wayside control system, the Genisys II non-vital logic emulators and 60 Hz track circuits.

The following year, CSX awarded a contract to Union Switch & Signal for Next Generation Dispatching Program at the Kenneth C. Dufford Transportation Center in Jacksonville, Florida. The program was designed to replace the dispatching system Union Switch & Signal originally installed in 1988. The Next Generation Dispatch program at CSX represented state-of-the-art technology in Centralized Traffic Control (CTC) and dispatching. The first two phases involved migrating the CTC and Dark Territory Control applications from the VAX-based system to the new Union Switch & Signal distributed, Unix-based platform. Subsequent phases of the project included the migration of the Train Message/Bulletin and Train Sheet functions.

The Union Pacific Harriman Dispatch Center in Omaha, Nebraska, controls freight and passenger traffic in the Central and Western United States.

The infrastructure business strong, Union Switch & Signal engineers continued to develop new products, including the first Electronic Circuit Controller (ECC) for its style "M" switch machines. The ECC replaced moving parts with an entirely self-contained electronics package, which detected and reported the position of the switch points, as well as evolving mechanical problem with the point assembly well before safety margins were exceeded. Among the first large-scale installations in which Union Switch & Signal introduced the ECC as part of its new M-23E electric switch machine was on the Charlotte Area Transit System (CATS) South Corridor Light Rail Project, which went into revenue service in November 2007.

Union Switch & Signal and other signaling industry competitors soon recognized the potential value of Light Emitting Diode (LED) signals, primarily for their ability to extend signal life and consume less wayside power than their incandescent predecessors. LED lights were best known as miniature, BB-sized indicators on electronics control panels and other user devices prior to their foray into the rail industry products. Industry advances in size, intensity and arraying made LEDs feasible to use as railway signals.

The CATS in Charlotte, North Carolina, is a double track light rail that covers 9.4 miles in Mecklenburg County. The Park and Ride system and buses feed into the 15 stations. "When we first opened," said John Bisser, rail systems manager for CATS, "they predicted we'd have 9,100 riders a day for the first year. We currently average 12,900 daily. It avoids downtown parking, which is minimal and expensive. It's quicker to get in and out of downtown if you go to an event. It stops at the Time Warner Cable Arena for concerts and sporting events, including Bobcats basketball games, and the Carolina Panthers Football play at Bank of America stadium, which is just five blocks from the nearest station."

Union Switch & Signal Project Manager C. J. Welter and a group of engineers including Mark Gruber (the wayside engineer) worked on the project that installed 28 M-23E switches on the main line, and one GE 3000 that exits the South yard lead. The line has 36 interlocking signals, 21 railroad crossing gates and hand throw switches in the yard. At the North yard lead – a section of rail for entering and exiting the yard from main line – a M-23 switch machine is connected to a derail which was designed such that if a train's brakes fail on the yard lead and the train moves toward the main line unexpectedly, the train would safely derail before reaching the mainline. Union Switch & Signal's AF-800N audio frequency train detection, wayside cab signaling system and a Union Switch & Signal central office System Control and Data Acquisition (SCADA) system completed the original installation.

In 2011, CATS trains transport 12,900 passengers daily with the Union Switch & Signal N2 transit dwarf signals (foreground) lighting the way.
Photo credit: Joanne L. Harris

Union Switch & Signal installed 21 railroad crossing gates at CATS.
Photo credit: Joanne L. Harris

In 2006, Union Switch & Signal incorporated the new LED technology into its LED color light signals, which were first deployed on a full-scale basis for the CATS project. Additional equipment required by incandescent-lamp signals, such as light-out relays and slide-wire resistors, were replaced with a simple set of electronic modules including a current regulator and several transient voltage protection boards. LED signals were projected to last up to 10 years before failure, far longer than the best incandescent lamp signals.

Other railroads saw the advantages of the LED signals. In 2011, the Buckingham Branch Railroad in Charlottesville, Virginia, began the task of replacing 110 signals in 70 locations along 125 miles of signaled territory from Orange to Clifton Forge, Virginia. The shortline owns 17 miles of track and leases about 280 more from CSX and Norfolk Southern. Union Switch & Signal installed color light signals back in the 1930s or 1940s, and they were all still functioning, but the signal controls were sent through old open wires on dilapidated pole lines. The railroad's Manager of Signals, Terry Wildermuth, said the old lines had lost their insulation over the years, which caused excessive red signals and unreliable service.

Union Switch & Signal engineers continued to branch out into new technologies. They developed a software program that provided real-time planning so that the railroads' central control office could safely increase the speed of trains passing along their rails. This software, the Optimizing Traffic Planner™ (OTP), dispatches trains according to a set of key business goals such as scheduling and routing, monitors the outcome of the goals in real-time and automatically adjusts subsequent dispatching decisions if the objectives are not met.

While these cutting edge technologies were honed, Union Switch & Signal continued to improve upon their bread and butter signaling system products. In the mid-2000s, the MicroTrax was converted from a standalone product into a component of the MicroLok II controller. MicroLok was designed as an interlocking controller, but MicroLok II additionally provided control for track circuits and the wayside elements of cab signaling. It also had the capability to communicate with the central office. Essentially, MicroLok II absorbed the software functions of Genisys and MicroTrax, and was now doing the job of three predecessor products.

Union Switch & Signal had another groundbreaking opportunity to create a prototype tool for multiple railroads to share information with each other within the spoke and hub-designed tracks of Chicago, Illinois, home to one of the most complex and congested territories in North America. In 2008, ten major railroad and passenger lines ran 1,300 trains

**"The changing of the guard"
Buckingham Branch
Railroad in Charlottesville,
Virginia, are replacing old
Union Switch & Signal
color light signals with
new LED signals.**

Photo Credit: Greg Gooden, Lead Signal Maintainer on the BBRR

Dr. Alan E. Calegari

The Last Union Switch & Signal President

By the time Alan Calegari was 19, he had lived and traveled in three continents and was back in Rome with his family. The young Calegari attended a Jesuit boarding school in Rome, making frequent trips to the United States to visit friends and family.

His Bachelor of Arts degree in Architecture "gave him a global view instead of a technical narrow view." He holds a Doctorate in Management from the University of Studies in Rome-Loyola University Foreign Exchange Program, attended an executive program at Harvard University and is a retired United States Marine Corps aviator officer.

He had a successful executive history with Pan Am World Services, Johnson Controls and Siemens Building Technologies before he was hired as President and CEO for Americas Region at Dedicated Micros, a British-owned security industry company.

He spent his free time in the late 1970s riding quarter horses, competing in California's regional and state jumping competitions and playing polo.

He picked up the game of tennis working at Pan Am World Services (PAWS), and was soon playing in tournaments on the bases where he worked. He then moved into the art of building wired and radio-controlled model airplanes.

"The difference between people who are 'widely successful' and those who are 'successful' is to figure out what questions are worth asking in the business world," explained Calegari. "That allows you to become customer centered and people centered. The success in the turnarounds I've done at Siemens, Dedicated Micros and here isn't because I'm a better businessperson. It's because I have the ability to prioritize what needs to happen, to understand the issues, to relate to colleagues and employees and to help make the change. It boils down to four things: Listen, learn, help and lead. You have to listen to what the problems are, learn about the issues and how to fix them, help the people to do it and lead them by setting the pace. You will never have everyone in sync with you. It's really invest, invest, invest – time, leadership, understanding and listening. It isn't just a financial goal."

His first turnaround was a project with PAWS at the Kings Bay, Georgia, nuclear submarine base. "We were financially losing and creating issue with our Navy customer. I stabilized the project, turned the financial situation around and rebuilt trust with the customer. We established an environment of mutual respect and total cooperation rather than finger pointing. This experience gave me a feeling for the passion of a turnaround. Taking a company from struggling to success was what I felt good at doing. My inspiring and most revered boss (Mr. Simpson) at the time counseled me to continue what I was doing and while doing it, to "make it better every time. You have the stuff to do these things," he said. I began talking with the customers and then going back to the team and getting their views on the situation. Customers always provide you with the best insight of the issues if you are willing to listen. You have to break the paradigm, establish the priorities, and address them with perfection. In doing so, you can establish lasting partnerships. One must always be ready to execute on the principle of "good enough. Issues must be confronted with ingenuity that sparks innovation, creativity and a global mindset."

> **"Genuine common sense is welcome in every business environment."**
> — Alan Calegari

Today, Alan Calegari spends what free time he has with his family. "I love to spend time with my family. I believe in the family environment and the warmth and support it can give you. We like to travel, discover cities, we go out for dinner and this keeps us connected. Then we do what we like to do – my wife and the girls enjoy shopping, theater, and what the local environment may provide for leisure and amusement. I like art, architecture and reading. I also like to work out in the yard, raking leaves and pretending to know what I am doing in pruning. Then my wife comes out and asks me what I'm doing. She reminds me that the landscaper is coming tomorrow."

daily though the Chicago area. Each railroad monitored its own right-of-way, but they relied on phone calls or faxes to facilitate train movement as they could not see the condition of the others' territories. The Chicago Transportation Coordination Office (CTCO), established in 1999, facilitated closer dispatch coordination between the passenger line and freight railroads, but communication problems still existed. Three groups – a collection of participating railroads represented by the Association of American Railroads (AAR), the Illinois Department of Transportation (IDOT), and the Chicago Department of Transportation (CDOT) – joined forces to establish the Chicago Region Environmental And Transportation Efficiency (CREATE) Program.

The COP system encompasses the overview display which depicts tracks, trains, movement authorities and field devices for the railroads involved; the CAD systems that interface to the Web Server; and instant messaging/report generation capabilities.

Union Switch & Signal approached CREATE with a prototype concept to improve the flow of railroad and highway traffic, reduce waiting times at highway crossings and reduce grade crossing hazards in the Chicago metropolitan area. Four railroads – BNSF, CSX, Indiana Harbor Belt (IHB) and UPRR – joined the pilot program to test the Chicago Common Operational Picture (COP) tool. COP Phase 1, completed in March 2008, provided a successful integration of the four railroads' dispatch displays and an open interface technology to enable all train, track and switch data from each participating dispatch system to be integrated into a single, real-time monitoring screen. "We haven't all been completely integrated to use the system adequately," said Scott Kuhner, superintendent at CSX, "but it is allowing us to make better moves, plan better for good train meets and have more efficient train moves."

The Railroad Research Foundation (RRF) received an additional grant from the FRA in 2009 to extend the prototype into a full-capacity system, which they awarded to Ansaldo STS USA. When fully integrated, the overview display system will encompass the entire Chicago area for the railroads that participate in the CREATE Program, merging the individual overview displays into a single, consolidated display that represents an area of approximately 20 by 20 miles. Jeff Harris, Assistant Vice President Operations Planning at Norfolk Southern, said that there is a two-fold reward. "COP gives you the opportunity to look at your trains moving on someone else's railroad. If I see someone else has a train moving in my direction, I can accept that train and keep the traffic moving. This is unique to any terminal in the country, for one railroad to see every other railroad. It has never been done before."

The COP system encompasses the overview display which depicts tracks, trains, movement authorities and field devices for the railroads involved; the Computer-Aided Dispatching (CAD) systems that interface to the Web Server; and instant messaging/report generation capabilities. "The COP will become more of an oversight tool for the handoffs between carriers," said Dave Grewe, director of UP and director of the CTCO in Chicago. "It's also a tool for crew utilization. You can track your own trains through the terminal, anticipate their arrival and plan crews appropriately."

America has experienced an endless cycle of bearish and bullish economies throughout its history, and the 2000s were no exception. In 2006, the economic bubble burst in the United States, as well as in other countries around the globe. Real estate values plunged, gas prices skyrocketed, and investors lost their shirts. In spite of the sudden recession, the rail industry was not hurt as heavily as other market segments, according to Steve Bolte, publisher of Progressive Railroading magazine. "The rail industry really didn't tank like the rest of the economy. The mechanical side didn't crash until about mid-2008," he recalled. "Then the brakes came on. Nobody was buying freight cars or locomotives. The railroads were mothballing a lot of their fleet because of the lack of freight business. That slowed that segment of the business. But the signal communication and maintenance of way investments stayed fairly strong."

In October 2008, Congress passed The Rail Safety Improvement Act of 2008 in response to the Sept 13, 2008, train crash in Chatsworth, California that killed 25 people. Alan Calegari had just become President and CEO of Union Switch & Signal in June.

"After the law passed, the FRA was mandated to enforce the application of Positive Train Control – PTC – and to define the cardinal rules," Alan Calegari explained. "The project would have to use mainly private funding, with the exception of some federal funding available for those passenger commuter lines that had freight trains encroaching upon their tracks. The freight lines looked at their tracks, crossings, and infrastructure and quickly assessed the value of a major investment to support the effectiveness of a PTC implementation."

The Rail Safety Improvement Act mandated that PTC equipment must be installed by the end of 2015 on all railroad main lines where there is track with a mix of passenger and hazardous materials. According to the FRA, railroads are currently looking at costs of up to $13.2 billion (in 2009 dollars) to install and maintain PTC over the next 20 years, but PTC will yield just $1 in benefits for every $20 spent on it.

"The PTC craze kept the market going," Steve Bolte observed. "Even though the volume was down, they still had to maintain their infrastructure. They still had to replace equipment, gates and crossings, regardless of how much traffic was being moved." Bolte stated that the PTC that drove the market at the end of the decade continues to do so at the present time. Railroads are developing systems and starting to implement some of the wayside equipment necessary to support the PTC technology. "There are a lot of upgrades in infrastructure, and new infrastructure, that have to be done. Radio control and GPS technology has to be put in place." In fact, radio towers, the foundations to support the towers,

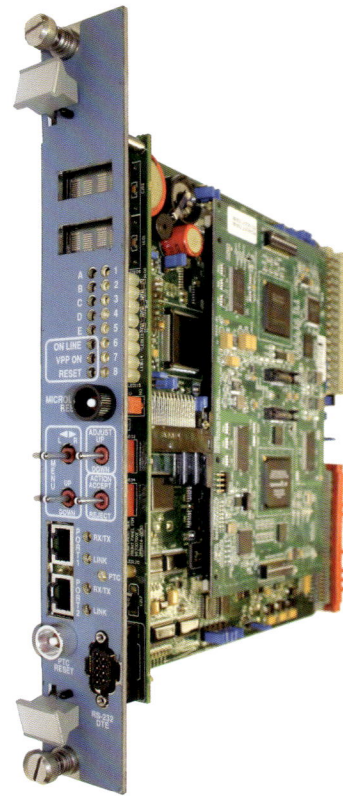

The VitalNet Current Sensor is a pre-wired unit that allows railroads to install PTC on lamp signals without removing any existing wires.

Current Positive Train Control, or PTC, product development includes the VitalNet™ CPU, a board upgrade that makes MicroLok II PTC-compatible (top), and the VitalNet Point Monitor, that continuously monitors the switch sensor's operation state and position in a vital (failsafe) manner (bottom).

The Staggers Act of 1980 proved to be a success for both railroads and shippers. After decades of steady decline, rail market share (measured in ton-miles) has trended slowly upward since the Act was passed. It is currently around 43 percent. Railroads are stronger financially. Return on net investment, which had been falling for decades, rose to 4.4 percent in the 1980s, 7.0 percent in the 1990s, and 8.0 percent from 2000 to 2008. Nevertheless, even in recent years, when railroads have had record traffic and earnings, the industry's profitability has been no better than average, and usually below average, among all industries.

antennas and bandwidth are all on the shopping list, in spite of the recession, which was still ongoing in 2011.

While the freight railroads in the Unites States push to meet the deadlines of PTC implementation, railroad and mass transit authorities around the world – particularly in the major up-and-coming economies of China and Brazil – continue to develop and expand their existing networks to accommodate an ever-increasing demand for transportation of freight cargo and passengers. "We have the large market share of mass transit in China," said Calegari. "With our local technology partner Insigma, we have secured simultaneously five major projects of major metro systems in China's fastest growing cities. At the same time, our parent company of Ansaldo STS in China secured a high-speed line program. We have nearly the exclusive driverless mass transit market share since the commissioning of the Copenhagen Metro. From our offices in the United States we provided engineering services for the driverless metro systems of Brescia, Rome and Milan. We are currently installing a driverless system in Riyadh, Kingdom of Saudi Arabia, for the University for Women."

Ansaldo STS is installing the venerable MicroLok II technology for the city of Sao Paulo Metro Transit Authority, CPTM (Companhia Paulista de Trens Metropolitanos) in Brazil. Both Brazil and Mexico are following the example of the United States by expanding railroad freight transportation. The United States maintains the best and largest rail freight system and that is why Ansaldo STS has always retained a strong market presence there.

"The railroads want to continuously improve their operations by maximizing every aspect of the train movement," said Calegari, "increasing in a safe way the velocity of their trains and the level of prioritization through their dispatching systems. Increasing velocity at pace with safety – moving all kind of freight in a more efficient, cost effective way – is the new paradigm. As railroad suppliers, we needed to address this new challenge."

On January 1, 2009, the Union Switch & Signal name and esteemed "pretzel" logo were retired from use and replaced by the Ansaldo STS-USA name and logo as the company progressed into the full integration of the company. The Ansaldo logo is rooted in the dynamic symbol of the Finmeccanica Group. Ansaldo, once the holding company for its global subsidiaries, has now integrated its companies in Italy, France, Spain, Portugal, UK, Australia, and the United States to eliminate conflicting business and to create a product portfolio for the best-of-breed and streamlines technology applications for projects around the world.

"Today," said Dr. Alan Calegari, "we are a group that can pride itself from having the best of products and technologies to address the requirements of any market economies and customers. In the United

States, we are both a local and a global company with strong components and systems-based business. Through our head offices and technology center in Pittsburgh, Pennsylvania, our yard business center of Norristown, Pennsylvania, our Engineering support centers of Hamilton, New Jersey, Jacksonville, Florida and manufacturing facility of Batesburg, South Carolina we are a significant presence on the market and continue to be an authoritative competence in the transportation industry. Through these locations we are currently installing among many others, new lines on WMATA in Washington, DC, PATH and LIRR in New York, Toronto Transit Authority in Canada, and completing a project for the Port Authority Allegheny County."

The Copenhagen Metro is a fully automated, driverless system that runs on Union Switch & Signal's Automatic Train Control (ATC) equipment, regulating speed control, routes trains and stops trains at stations. Within one month of opening, the Metro served over one million riders.

171

The human goal is to provide safety on the rails. Ansaldo STS engineers are as creative today as their counterparts were 130 year ago. They are constantly looking for ways to meet the ever-increasing demands of railroads and metros around the world for enhancing not only safety, but also operational efficiencies, speed and capacity. They are using all new approaches for safety-critical systems, optimized freight rail traffic planning for entire Class 1 railroads, and radio communications-based train control for complex mass transit systems. These advanced technologies would astound even George Westinghouse…or perhaps not. Perhaps his company is right where he would expect it to be.

AnsaldoSTS

A Finmeccanica Company

NEWS

1881: Westinghouse Forms Union Switch & Signal

Business &

NEWS

Dateline: January 1, 2009
Union Switch & Signal Becomes Ansaldo STS USA

ry 1, 2009 · Union Switch & Signal Becomes A

Ansal

The business that George Westinghouse founded in 1881 continues to thrive, as it has become a fully integrated part of Ansaldo STS.

Hall Of Fame

There are never enough pages in a book to cover the achievements of the thousands of dedicated employees of a company over the course of 130 years. To do justice to every significant contributor to the organization would surely result in a multi-volume series. Throughout the research process, the stories told by those still living mentioned the names of "those who could not be left out." And yet, there remained insufficient space to applaud the achievements of so many people in such a short amount of space.

To that end, a short list of people was cultivated from the names that continued to surface as the book progressed. The list is not exhaustive. It merely serves to remind us that while world and local events have caused change in the rail industry, and products are invented because of that change, the ultimate ingredient that determines a company's success or failure is its people.

It is the person who is not afraid to fail who will succeed. It is the employee who donates time to community service that makes them whole people. And it is the person with "a better idea" for safety on the rails that continues to make Ansaldo STS USA the world leader that it is today. The journey of Union Switch & Signal has taken many turns throughout its 130 years of history, and with proper leadership and the dedication to succeed, the company will last another 130 years

Andrew "Andy" J. Carey

Andy Carey entered the Union Switch & Signal apprentice program in 1949 after graduating from the University of Pittsburgh with Bachelor of Science degree in Electrical Engineering. Apprentices rotated through five engineering groups. When he completed the rotation, he chose to work in track circuits under Crawford "Tacky" Staples, and worked under him for about 15 years. (The apprenticeship program ended by the 1960s.)

From a supervisor on the track circuit division through the ranks to be the head of R&D, Carey watched the company transition from a hardware company to a software company, and was awarded two patents along the way. He went on to become Vice President of Engineering, and headed the construction company that installed the equipment. Andy Carey ended his career in Special Projects and retired in 1994, having been "always the first man in the plant and the last man out."

Carey is remembered as being a nice guy, but it was not simply pleasantries – he was ethical and kind to people. He broke the "taboo" by bringing capable women into production meetings, and gave people career opportunities that they would have otherwise never been afforded. His belief was, "You can't bully people into doing something; you have to get their respect."

Joseph "Rocky" D. Eiseman

Rocky Eiseman started working in 1973 at the Swissvale plant in the stock room and the shipping department. His hard work was rewarded and he moved into the Customer service, Planning and Materials department and became Supervisor over Materials. He moved to Augusta, Georgia, to head the Customer Service Center and Training Center. Just two years later, Eiseman became Manager of the Augusta facility.

He returned to Batesburg in 1989 as Manager of Shop Operations, was promoted to Plant Manager and currently serves as Vice President of Manufacturing and Supply Chain.

Many plants attempt to increase productivity by hiring outside consultants to tell them how to make their plants run more efficiently. But in 2009, Rocky Eiseman approached his plant employees to address the challenges the shops had with manufacturing. He had the wisdom to realize that the people doing the work knew what type of shop layout would best suit the people who had to move parts from one place to another. Eiseman oversaw the team effort as they worked to implement a program called "Lean Manufacturing." The shops were reconfigured to make the process flow more smoothly, which resulted in a component operation general profit increase of 12 percent within a year.

Dorothy Kolano

In June 1944, Dorothy Kolano had to get a work permit to take a job at Union Switch & Signal at age 17. Her first job was in the Orders Department as a secretary. She moved through a series of departments over the years as she proved her value to the company's Engineering, Research, Marketing, and Quality Control personnel. She worked for engineers and then for several Vice Presidents. Secretarial services were always in demand at the company, and she felt the women were treated well and with respect.

Kolano started in the era of the manual typewriter, and as secretaries did those days, she took dictation – a skill that today is nearly a lost art. As time and technology progressed, she secured an electric typewriter. She remembers that in those days, they had carbon copies if someone wanted more than one copy, and she thought the Xerox machine was great. Dorothy Kolano spent her entire career working at "the Switch" – for 50 years, before finally retiring in 1994, shortly before the company made its move into the current facility on Technology Drive.

Of all the projects she saw go through the engineering department, she felt that the Automatic Car Identification (ACI) system of the 1960s was one product she wished they had held onto, as it was a predecessor for modern day bar code technology.

Darde Khan

Darde Khan started her career in 1993 as a Technical Writer composing documentation for the company's central office systems department. She currently manages the Training and Documentation (T&D) Group. Because she reviews every project that comes through the company, she has met nearly everyone in the building. Such an expansive scope of the company's business has given her a broad perspective of the company both domestically and internationally. And yet, she represents the hundreds of employees who contribute so greatly to the company without the prestige of patents or international awards, but without whom the company would suffer.

The T&D group writes all of the Operations and Maintenance Manuals, material and training plans and Darde prepares every T&D systems estimate. Under Darde's guidance, the T&D Group is maturing into a department that performs in-the-field training classes for customers, as well as produces training documentation.

Being familiar with another male-dominated business – the industry of boating – she compares the ability for women to come into their own in the railroad business. Technical writing has opened up the door for women who want to rise to the challenge of learning and working in the industry.

Robert "Bo" Malackany

In July 1952, right out of high school, Bo Malackany took his first job working at Union Switch & Signal in the Swissvale storeroom as an attendant. After five years he transferred into the Data Processing department where he worked for 27 years. These were the days of the IBM key punch cards. The first computer Malackany worked on was a colossal unit with one- to four-part printers. He worked mostly as the printer operator, or Tabulation Machine Operator, running a "burster" that broke all the sheets (perforated printouts), and then put the sheets into books.

During a layoff, Malakany took a "bump" into the mailroom as the shop and office shared the same union. He continued to work in the mailroom as a Senior Office Clerk with one helper, until the company came to him and said that they needed to cut the mailroom down to one person. Bo Malakany was in charge of the mailroom at the time, with 47 years and nine months of service. He knew his helper needed the job, and Malakany was already at retirement age. He gave up his opportunity for a 50-year service award and took the buyout in April of 2000 so his helper could keep his job.

The mailroom work was outsourced by the end of the year, and Malakany was rehired into the outside company. Eleven years later, he is still on the job, along with his pal "Ro Bo," the robotic mail carrier system that someone anonymously named after "Bo" the first day it went into service.

Chinnarao Mokkapati

Chinnarao Mokkapati was an assistant professor at the Regional Engineering College in Warangal, India in the mid 1970s. Wanting to obtain his PhD, he came to America to study at West Virginia University (WVU) in Morgantown, West Virginia where he planned to obtain his PhD in Electrical Engineering, then return to his homeland. While working on a WVU project to improve mine power system safety (funded by the US Bureau of Mines), he was discovered by Union Switch & Signal. System safety was a new field in the late 1970s, and the company hired him in 1979 "before he could say no."

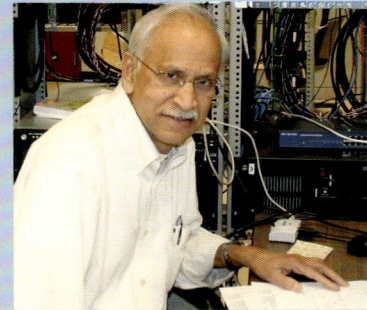

The company had several projects underway on the Baltimore and Miami transit lines and the Northeast Corridor, all of which required system safety experts. These projects stipulated safety and reliability studies and testing, which were new in those days and critical for eliminating any and all hazards during train movements.

Mokkapati progressed from an Engineer "A" to manager and director positions, and in time became Vice President of Quality & Systems Assurance. In 2011, with multiple white papers published in an array of professional journals and several patents to his credit, he serves as Chief Technologist - Systems Assurance.

Chinnaro Mokkapati oversaw the independent safety certification of the world's first Driverless Automatic Train Control System for the Copenhagen Metro, in compliance with European CENELEC standards, and he continues to contribute to the design, testing and safety certification of Positive Train Control Systems (PTC) for railroads and Communication Based Train Control Systems (CBTC) for metros. Sharing his experience with others, he trains and mentors engineers in the area of system safety, reliability and electromagnetic compatibility assurance.

Denny Pascoe

Robert D. ("Denny") Pascoe remains one of the rail industry's most respected and sought-after train control experts, even after his retirement from Ansaldo STS USA in 2010. Since starting his career at Union Switch & Signal in 1975, Denny has earned the respect of his colleagues throughout the company, the industry and government for his unmatched genius in the field of train control safety systems. His dedication to his field culminated in the Copenhagen Driverless Automatic Train Control System, the first and only such system to meet rigid European passenger safety standards.

The demand for Denny's expertise is reflected in the numerous professional papers and proposals he has prepared for the Federal Railroad Administration, the Federal Transit Administration and their companion organizations outside of government. As such, he has become a natural choice as an Expert Witness in federal rail safety investigations. In all of his dealings inside and outside the company, Denny is well known for his humility, approachability and undivided love for his profession. Not surprisingly, he turned down several offers to lead large engineering departments so that he devote full time to the latest engineering frontiers.

At his retirement dinner in late 2010, Denny the ever-humble Mr. Pascoe stated that "I was privileged to stand on the shoulders of many great people at Union Switch & Signal."

Glossary

AAR	American Association of Railroads	ICC	Interstate Commerce Commission
ACI	Automatic Car Identification	IDOT	Illinois Department of Transportation
ACIU	Allegheny Congenial Industrial Union	IHB	Indiana Harbor Belt
AF	Audio frequency	ITC	Inductive Train Communication
AFO	Audio Frequency Overlay	L&M	Liverpool and Manchester
AIEE	American Institute for Electrical Engineers	LACMTA	Los Angeles County Metropolitan Transportation Authority
ARR	Alaska Railroad		
ATC	Automatic Train Control	LED	Light Emitting Diode
ATG	Advanced Technology Group	LIRR	Long Island Railroad
ATO	Automatic Train Operation	MBTA	Massachusetts Bay Transportation Authority
ATP	Automatic Train Protection	MNCR	Metro-North Commuter Railroad
B&O	Baltimore & Ohio	NIST	National Institute of Standards and Technology
BART	Bay Area Rapid Transit District	NJT	New Jersey Transit
BNSF	Burlington Northern Santa Fe Railway	NRPC	National Railroad Passenger Corporation
C&O	Chesapeake & Ohio	OPEC	Organization of Petroleum Exporting Countries
CAD	Computer-Aided Dispatching	OTP	Optimizing Traffic Planner™
CATS	Charlotte Area Transit System	P&W	Providence & Worcester
CBTC	Communications-Based Train Control	PAAC	Port Authority of Allegheny County
CDOT	Chicago Department of Transportation	PAT	Port Authority Transit
CEO	Chief Executive Officer	PATH	Port Authority Trans-Hudson
COP	Chicago Common Operational Picture	PRR	Pennsylvania Railroad
CREATE	Chicago Region Environmental And Transportation Efficiency	PTC	Positive Train Control
		RRF	Railroad Research Foundation
CRT	Cathode-Ray Tube	RRIF	Railroad Rehabilitation and Improvement Financing
CTA	Chicago Transit Authority		
CTC	Centralized Traffic Control	SCR	Signal Corps Radar
CTCO	Chicago Transportation Coordination Office	SEC	Securities and Exchange Commission
DART	Dallas Area Rapid Transit	SEPTA	Southeastern Pennsylvania Transportation Authority
DSL	Dynamic Sciences Limited		
ECC	Electronic Circuit Controller	TAC	Touch Activated Control
EG&G	Edgerton, Germeshausent and Grier	WABCO	Westinghouse Air Brake Company
FRA	Federal Railroad Administration	WMATA	Washington Metropolitan Area Transit Authority

Illustration Sources

Picture Credits

1. Great Barrington Historical Society Collection: 37

2. Library and Archives Division, Sen. John Heinz History Center: 99, 100, 101, 102

3. Library of Congress: 6, 7, 13, 29, 35, 41, 42, 50, 53, 60, 66, 68, 69

4. Pennsylvania Historical and Museum Commission and the Railroad Museum of Pennsylvania: Forward, 48, 53, 60, 90, 95, 97, 104, 105, 108, 109, 110, 111, 128

5. Railway Signaling: 51

6. Union Switch & Signal, Ansaldo STS Collection: 18, 28, 33, 53, 62, 78, 83, 88, 89, 90, 97, 101, 111, 114 117, 119, 128, 129, 133, 136, 138, 139, 140, 142, 143, 144, 147, 149, 148, 150, 152, 153, 154, 155, 160, 161, 162, 163, 164, 166, 168, 169, 170, 171

7. Union Switch and Signal Strike Photograph Collection, June 1914 AIS.1991.03 Archives Service Center University of Pittsburgh: 64

Special Thanks

Ben Feely

David Grinnell, Senator John Heinz History Center

Lora Hearn, Wilmerding Renewed Incorporated (Westinghouse Castle)

Geraldine Homitz, Wilmerding Renewed Incorporated (Westinghouse Castle)

Cyrus Hosmer, III

Arthur Humphrey, Great-Grandson of Arthur L. Humphrey

Bradley Smith, Railroad Museum of Pennsylvania

Kathleen Wendell, Senator John Heinz History Center

Nick Zmijewski, Railroad Museum of Pennsylvania

Additional Acknowledgements

George W. Baughman III

John Bisser

Stephen Bodnar

Stephen Bolte

John G. "Jack" Borman

Buckingham Branch Railroad

Andrew Carey

Charlotte Area Transit System

Joseph Eiseman

Virginia Farnsworth

Raymond Franke

Zachary Gillihan

David R. Grewe

Barbara Henninger

Frank Himmler

Jim Kovach

Mary Beth Kovic

Thomas P. Lawton

Ed Meek

Denny Pascoe

Railroad Museum of Pennsylvania

Jon R. Roma

Alberto Rosania

Gary Ryker

Senator John Heinz History Center

Scott Sluis

Kathryn Spear

Glenn Stinson

Arthur Ticknor

Rich White, II

Union Pacific

Wilmerding Renewed Incorporated (Westinghouse Castle)

Bibliography

Wilson, H. R. *Railway Signaling*. London, 1900. Print.

Spang, Henry W. *The Treatise to Perfect Railway Signaling*. 1902. Print.

Brignano, Mary, and Hax McCullough. *The Search for Safety: A History of Railroad Signals and the People Who Made Them*. American Standard, 1981. Print.

Martin, Albro. *Railroads Triumphant: The Growth, Rejection, and Rebirth of a Vital American Force*. Oxford UP, 1992. Print.

American Railway Association, Signal Section. *The Invention of the Track Circuit*. American Railway Association, Signal Section, 1922. Print.

The Union Switch and Signal Co. *Pocket Reference Book*. 1889. Print.

Garbedian, Gordon H. *George Westinghouse*. New York: Dodd, Mead &, 1943. Print.

Solomon, Brian. *Railroad Signaling*. Minneapolis: Voyageur, 2010. Print.

Spang, Henry W. *The Treatise to Perfect Railway Signaling*. 1902. Print.

Cornwell, E. L. *The Pictorial Story of Railways*. Crescent, 1972. Print.

Levine, I. E. *Inventive Wizard: George Westinghouse*. New York: Julian Messner, 1962. Print.

Killikelly, Sarah H. *The History of Pittsburgh: Its Rise and Progress*. B.C. & Gordon Montgomery, 1906. Print.

Marshall, John. *Rail Facts and Feats*. New York: Two Continents Group, 1974. Print.

Carter, Ernest F. *The Railroad Encyclopedia*. London, 1963. Print.

Crane, Frank. *George Westinghouse: His Life and Achievements*. New York: WM. H. Wise &, 1925. Print.

Nock, O. S. *Railways Then and Now: A World History*. New York: Crown, 1975. Print.

The Union Switch and Signal Co. *Pocket Reference Book*. 1889. Print.

American Railway Association, Signal Section. *The Invention of the Track Circuit*. American Railway Association, Signal Section, 1922. Print.

Prout, Henry Goslee. *A Life of George Westinghouse*. C. Scribner, 1922. Print.

Stover, John F. *American Railroads*. 2nd ed. Chicago: University of Chicago, 1997. Print.

Wilson, H. R. *Railway Signaling*. London, 1900. Print.

Balliet, Herbert S., Keith E. Kellenberger, and Henry M. Sperry. "Dr. Robinson's Record from Wesleyan University." *The Invention of the Track Circuit*. New York: Signal Section, American Railway Association, 1922. 58-67. *Internet Archive*. 31 Jan. 2007. Web. 28 Jan. 2011.

Association of Engineering Societies (U.S.). "Proposed Improvements in St. Louis Terminals." *Journal of the Association of Engineering Societies*. Vol. 32. Board of Managers, 1904. 23-24. *Google Books*. 07 May 2009. Web. 11 Feb. 2011.

American Society of Mechanical Engineers. "Gas Power Plants." *The Engineering Index Annual for 1909*. American Society of Mechanical Engineers, 1910. 217. *Google Books*. 11 June 2007. Web. 18 Feb. 2011.

Shepley, Carol Ferring. "Guido Pantaleoni." *Movers and Shakers,*

Scalawags and Suffragettes: Tales from Bellefontaine Cemetery. Illustrated ed. Missouri History Museum, 2008. 240. *Google Books*. Web. 10 Feb. 2011.

"Obituary." *Railway Signaling and Communications*. Vol. 12. Simmons-Boardman, 1919. 69-70. *Google Books*. 26 Oct. 2009. Web. 11 Feb. 2011.

Spicer, V. K. "Pro Sig." *Electric Locking, Ancient and Modern*. Vol. II. 358. Print.

"Westinghouse Companies Exhibits at the International Railway Congress, Washington, Nineteen Hundred and Five : Free Download & Streaming : Internet Archive." *The Westinghouse Companies Exhibits at the International Railway Congress, Washington, Nineteen Hundred and Five*. Wilmerding: Westinghouse Air Brake, 1906. *Internet Archive*. 18 Sept. 2007. Web. 11 Feb. 2011.

"Record of Electrical Patents." *Electrical Review and Western Electrician with Which Is Consolidated Electrocraft*. Vol. 64. Electrical Review Pub., 1914. 359+. *Google Books*. 6 May 2010. Web. 1 Mar. 2011.

Vantuono, William C. "The Supplier Side of Movement Planning. | Railway Age | Find Articles at BNET." *Railway Age*. 2005. *BNET*. Web. 18 Mar. 2011.

Todd, Frank Morton, and Panama-Pacific International Exposition Company. "Chapter XXIX In Factory and Field." *The Story of the Exposition: Being the Official History of the International Celebration Held at San Francisco in 1915 to Commemorate the Discovery of the Pacific Ocean and the Construction of the Panama Canal*. G.P. Putnam's Sons, 1921. 168-69. *Google Books*. 08 Oct. 2007. Web. 28 Feb. 2011.

"Railway Signal Association." *Railway Age Gazette*. Vol. 53. Simmons-Boardman, 1912. 729. *Google Books*. 08 Jan. 2007. Web. 10 Feb. 2011.

Official Gazette of the United States Patent Office. Vol. 10. Office, 1887. 667. *Google Books*. 09 Sept. 2008. Web. 1 Feb. 2011.

Luepp. *George Westinghouse*. 102. Print.

"New Switch at Ansaldo Signal." *Railway Age*. Simmons-Boardman, 1997. *HighBeam*. Web. 22 Mar. 2011.

Cunningham, Joseph J. "Roots of an Evolution: Fifty Years of Automated Train Control in New York." *Railway Age*. Simmons-Boardman, 2009. *BNET*. Web. 28 Feb. 2011.

"Supply Trade News." *Railway Age Gazette*. Vol. 59. Simmons-Boardman, 1915. 714. *Google Books*. 9 Jan. 2007. Web. 1 Mar. 2011.

"Business Notes." *The National Engineer*. Vol. 11. National Association of Power Engineers, 1907. 39. *Google Books*. 12 July 2006. Web. 10 Feb. 2011.

Forney, Matthias Nace. "Car Mail." *The Car-builder's Dictionary*. Railway Gazette, 1881. 26. *Google Books*. 31 Mar. 2010. Web. 05 Feb. 2011.

"Industrial Notes." *The Engineering Magazine*. Vol. 28. Engineering Magazine, 1905. IV. *Google Books*. 12 Dec. 2008. Web. 21 Feb. 2011.

"W. H. Higgins." *Railway Signaling and Communications*. Vol. 12. Simmons-Boardman, 1919. 69. *Google Books*. 26 Oct. 2009. Web. 1 Mar. 2011.

Atwood, L. C. *Practical Dynamo Building, with Detail Drawings and Instructions for Winding*. St. Louis: Nixon-Jones, 1893. *Internet Archive*. 14 Mar. 2007. Web. 07 Feb. 2011.

"Cost of Stopping Trains - When It Is Cheaper to Install Interlocking Signals." *Engineering and Contracting*. Vol. 34. Gillette, 1910. 440. *Google Books*. 27 Oct. 2009. Web. 06 Feb. 2011.

"Cost of Stopping Trains, Compared with the Cost of Maintenance, Operation and Inspection of Interlocking Plants." *The Railway Age*. Vol. 40. 1905. 445. *Google Books*. 04 Feb. 2010. Web. 06 Feb. 2011.

Franklin Institute. "Mechanical and Engineering Section. Fuel Oil." *Journal of the Franklin Institute*. Vol. 156. Pergamon, 1903. 163-67. *Google Books*. 26 July 2007. Web. 31 Jan. 2011.

"UNION SWITCH & SIGNAL.(John Mandeli Named President)." *Railway Age*. Simmons-Boardman, 1998. *HighBeam*. Web. 22 Mar. 2011.

"1917 Fire Losses $267,273,140." *Quarterly of the National Fire Protection Association*. Vol. 11. National Fire Protection Association, 1917. 236. *Google Books*. 30 Sept. 2009. Web. 18 Feb. 2011.

Axelrod, Alan, and Charles Philips. "Standard Time Comes to America." *What Every American Should Know about American History: 225 Events That Shaped the Nation*. Adams Media, 2008. 174. *Google Books*. Web. 31 Jan. 2011.

Luepp. *George Westinghouse*. 102. Print.

"Manufacturing and Trade Notes." *Electrical Engineer*. Vol. 7. Electrical Engineer, 1888. 283. *Google Books*. 7 July 2006. Web. 08 Feb. 2011.

"Supply Trade News." *Railway Age Gazette*. Vol. 51. Simmons-Boardman, 1911. 104. *Google Books*. 8 Jan. 2007. Web. 15 Feb. 2011.

Schatz, Ronald W. "Chapter 2 The Workers." *The Electrical Workers: A History of Labor at General Electric and Westinghouse, 1923-60 The Working Class in American History*. University of Illinois, 1987. 37-38. *Google Books*. Web. 16 Feb. 2011.

Four-party. Vol. Bound. 103. Print.

"What Is Burning." *The Insurance Press*. Vol. 44. F. Webster, 1917. 110. *Google Books*. 9 Nov. 2009. Web. 18 Feb. 2011.

"Supply Trade Section." *Railway Age Gazette*. Vol. 48. Railway Age Gazette, 1910. 1019. *Google Books*. 05 Mar. 2010. Web. 18 Feb. 2011.

American Railway Association, Signal Section. *Proceedings*. Vol. VI. American Railway Association, Signal Section. 7. Print.

Blaise, Clark. "6 The Practice of Time." *Time Lord: Sir Sandford Fleming and the Creation of Standard Time*. Random House, 2002. 96-97. *Google Books*. Web. 31 Jan. 2011.

American Railway Association, Signal Section. *Proceedings*. Vol. VI. American Railway Association, Signal Section. 7. Print.

"The International Electrical Exhibition at Philadelphia." *The Electrician and Electrical Engineer*. Vol. 3. Electrical, 1884. 220. *Google Books*. 15 June 2006. Web. 7 Feb. 2011.

Kirby, Richard Shelton. "12 Modern Transportation." *Engineering in History*. Courier Dover Publications, 1990. 390. *Google Books*. Web. 26 Jan. 2011.

"Signal Supply." *Railway Signaling and Communications*. Vol. 10. Simmons-Boardman, 1917. 100. *Google Books*. 26 Oct. 2009. Web. 18 Feb. 2011.

Howard, L. F. "Track Circuit Signaling." *Trans*. AIEE, 1907. 1535. Print.

Kichenside, Geoffrey. "Chapter 8 - Stray Wagons and Breakaways." *Red for Danger*. 4th ed. Newton Abbot: David & Charles, 1982. 176. Print.

"Suppliers." *Railway Age*. 2009. 36. *Nxtbook*. Web. 22 Mar. 2011.

Judge, Tom. "Getting Wayside Data on Board: as Technology Improves, Communications between Train and Wayside Boost Ability to Forecast and Prevent Locomotive Failure and to Improve Safe, Efficient Train Handling - Industry Overview | Railway Age | Find Articles at BNET." *Railway Age*. 2003. *BNET*. Web. 18 Mar. 2011.

Spang, Henry W. "Development of Automatic Block Signaling." *A Treatise on Perfect Railway Signaling*. 1902. 60. *Google Books*. 6 July 2007. Web. 31 Jan. 2011.

Scientific American. Vol. 10. Munn &, 1864. 289+. *Google Books*. 4 Nov. 2009. Web. 05 Feb. 2011.

"The Swissvale Fire." *Municipal Journal*. Vol. 42. Municipal Journal and Engineer, 1917. 316. *Google Books*. 10 Mar. 2010. Web. 18 Feb. 2011.

"Electrical Engineers: William Stanley, Jr." *Electrical Engineer*. Vol. 10. Electrical Engineer, 1890. 179-80. *Google Books*. 1 June 2006. Web. 7 Feb. 2011.

"General News Department." *Railway Age Gazette*. Vol. 56. Simmons-Boardman, 1914. 1553. *Google Books*. 08 Jan. 2007. Web. 16 Feb. 2011.

Fleming, George Thornton, and American Historical Society. "Section 10." *History of Pittsburgh and Environs: from Prehistoric Days to the Beginning of the American Revolution*. Vol. 6. American Historical Society, 1922. 106. *Google Books*. 7 Sept. 2007. Web. 28 Feb. 2011.

Four-party. Vol. Bound. 200. Print.

Bianculli, Anthony J. "XVIII." *Trains and Technology The American Railroad in the Nineteenth Century*. Bridges and Tunnels Signals ed. Vol. 4. University of Delaware, 2003. 134. *Google Books*. Web. 26 Jan. 2011.

Spicer, V. K. "Pro Sig." *Electric Locking, Ancient and Modern*. Vol. II. 358. Print.

Westinghouse Air Brake Company. "Section 9." *Westinghouse Quick-action Automatic Brake Equipment*. Westinghouse Air Brake, 1920. 99. *Google Books*. 24 Aug. 2006. Web. 28 Feb. 2011.

"Signal and Interlocking." *The Railway Age*. Vol. 42. 1906. 731. *Google Books*. 7 Apr. 2010. Web. 10 Feb. 2011.

Skrabec, Quentin R., and Quentin R. Skrabec Jr. "Chapter 6: American and International Industrialist." *George Westinghouse: Gentle Genius*. Algora, 2007. 83-84. *Google Books*. Web. 14 Feb. 2011.

"The Plant of the Union Switch & Signal Co." *The Signal Engineer*. Vol. 2. 1909. 154-60. Print.

Proceedings of the American Railway Association. Emergency Session, Waldorf-Astoria Hotel, New York. 1907. 76. Print.

Proceedings of the American Railway Association. Special Session, Chicago. 1908. 270-71. Print.

Proceedings of the American Railway Association. Auditorium Hotel, Chicago. 1907. 27+. Print.

Electric RR Signal Co v. Hall Railway Signal Co. 6 Fed. 1881. Print.

Electric RR Signal Co v. Hall Railway Signal Co. 6 Fed. 1881. Print.

Westinghouse Air Brake Company v. The United States. United States Court of Claims. 12 Mar. 1965. Print.

Westinghouse Airbrake Company v. The United States. United States Court of Claims. 12 Mar. 1965. Print.

Reis, Ed. *Early Lamps by Westinghouse*. PDF.

Lewis, L. V. Interview. Print.

Telephone interview. 16 Mar. 2011.

Frothingham, E. L. Letter. 16 Apr. 1881. MS. 85 Devonshire Street,

Boston, Mass.

Frothingham, E. L. Letter. 14 Apr. 1881. MS. 5 Devonshire Street, Boston, Mass.

"W. W. Salmon." Letter. 10 Nov. 1927. MS.

The Winged Head Aug. 1945: 21. Print.

Railway Gazette 13 Oct. 1876: 443. Print.

The Bulletin Index, Section II 29 Apr. 1943. Print.

Railway Gazette 21 Dec. 1883: 845. Print.

Railway Signal Engineer 1922: 219. Print.

Railway Age 13 Feb. 1903: 202. Print.

Signal Engineer Apr. 1914: 104. Print.

Signal Engineer Aug. 1914. Print.

Railway Gazette 22 Nov. 1878: 569. Print.

Railway Signaling 1938: 83. Print.

"Biography of F. L. Pope." *Electrical Engineer* May 1934: 788. Print.

Railway Gazette 16 Oct. 1875: 424+. Print.

Signal Engineer Jan. 1916. Print.

Railway Signaling June 1929: 224. Print.

Railway Signal Engineer 1923: 400. Print.

Railway Signaling 1917: 32. Print.

Railway Gazette 1 June 1877. Print.

Railway Signaling 1936: 644. Print.

Telegraph & Telephone Age Dec. 1944: 19. Print.

Railroad Gazette 29 May 1905: 169. Print.

Railway Signal Engineer 1921: 437. Print.

Railroad Gazette 22 Nov. 1878: 569. Print.

Railway Gazette 22 Nov. 1878: 569. Print.

"Biography of F. L. Pope." *Electrical Engineer* May 1934: 788. Print.

Railway Signaling 1937: 306. Print.

Railway Gazette 22 Apr. 1881: 219. Print.

Railway Gazette 7 Nov. 1879: 592. Print.

Railway Gazette 8 June 1877. Print.

Railway Signaling June 1929: 224. Print.

Railway Gazette 12 Mar. 1880. Print.

Signal Engineer 1915: 360. Print.

Railway Signal Engineer 1925: 406. Print.

Railroad Gazette 2 Mar. 1883: 137. Print.

Railway Signaling 1928: 446. Print.

Signal Engineer Oct. 1909: 154-60. Print.

Railway Signaling 1927: 364. Print.

Signal Engineer 1915: 188. Print.

Railway Age 30 May 1891. Print.

Railway Review 22 Nov. 1890. Print.

Railway Gazette 16 Oct. 1875: 424+. Print.

Railway Gazette 18 Mar. 1881: 155. Print.

The Ohio State Engineer Nov. 1920: 23. Print.

Railway Gazette 13 Oct. 1876: 443. Print.

Signal Engineer 1914: 63. Print.

Railroad Gazette 2 Mar. 1883: 137. Print.

Railway Signaling 1918: 2. Print.

Railroad Gazette 27 Sept. 1907. Print.

Signal Engineer Oct. 1909: 154-60. Print.

Railroad Gazette 23 June 1879: 359. Print.

Signal Engineer 1916: 359. Print.

Railway Review 15 Feb. 1890. Print.

Railroad Gazette 12 May 1905: 155. Print.

Railroad Gazette 24 Dec. 1915. Print.

Railway Gazette 10 Jan. 1890. Print.

Railroad Gazette 23 June 1879: 359. Print.

Railway Age 13 Feb. 1903: 202. Print.

Railway Gazette 18 Mar. 1881: 155. Print.

Railroad Gazette 22 Nov. 1878: 569. Print.

Railway Gazette 1 June 1877. Print.

Railway Signal Engineer 1924: 221. Print.

Railway Gazette 25 May 1877. Print.

Railway Review 15 Feb. 1890. Print.

The Ohio State Engineer May 1925: 24. Print.

Signal Engineer Oct. 1909: 154-60. Print.

Railway Signaling 1925: 271. Print.

Railway Signal Engineer 1920: 30. Print.

Railway Gazette 1881: 166. Print.

Railroad Gazette 22 Nov. 1878: 569. Print.

Railway Gazette 13 Feb. 1880: 85+. Print.

Railway Signaling 1944: 640. Print.

Electrical Review 21 Mar. 1891. Print.

Railway Review 22 Nov. 1890. Print.

Railway Signaling 1926: 137. Print.

Railway Gazette 8 June 1877. Print.

Railroad Gazette 18 Feb. 1881: 103. Print.

Electrical Review 21 Mar. 1891. Print.

Railway Gazette 25 May 1877. Print.

Railway Gazette 3 Dec. 1880: 649+. Print.

"Coded Track-Circuit Signaling on the Pennsylvania." *Railway Signaling* May 1935: 245-52. Print.

Railroad Gazette 18 Feb. 1881: 103. Print.

National Car & Locomotive Builder June 1889. Print.

Railway Signaling 1944: 152. Print.

National Car & Locomotive Builder June 1889. Print.

Railway Gazette 21 Dec. 1883: 845. Print.

Railway Age 30 May 1891. Print.

Railway Gazette 10 Jan. 1890. Print.

Railroad Gazette 6 June 1884: 427. Print.

Railway Signaling 1926: 108. Print.

Railway Gazette 22 Apr. 1881: 219. Print.

Railway Gazette 13 Feb. 1880: 85+. Print.

"CORPORATIONS: Repeat Performance." *Time* 25 May 1953. *Time*. Web. 2 Mar. 2011.

Railway Gazette 12 Mar. 1880. Print.

Railway Gazette 7 Nov. 1879: 592. Print.

Railway Gazette 1881: 166. Print.

Railway Gazette 3 Dec. 1880: 649+. Print.

Railway Mechanical Engineer May 1946. Print.

Railroad Gazette 6 June 1884: 427. Print.

Stryker, C. E. "Charging Railway Signal Batteries." *Railway Signaling* 1925: 63. Print.

"21 Honored at Dinner in Batesburg." *The SWITCH* (Jan.-Feb. 1997): 7. Print.

Ticknor, Arthur W. "President's Corner." *The SWITCH* (Apr. 1989): 2. Print.

Doran, Clare E. "New Division Chief Says "Let's Keep It Moving!"" *Batesburg HI-LITES* (Aug.-Sept. 1980): 2. Print.

Alessandrini, Walter. "President's Corner." *The SWITCH* (Mar. 1990): 2. Print.

Ticknor, Arthur W. "President's Corner." *The SWITCH* (July 1990): 2. Print.

Doran, Clare E. "Highlights From the Past." *SwitchPoints* (Mar.-Apr. 1981): 8-11, 13. Print.

"Ryker Selected as President and CEO." *The SWITCH* (July-Aug. 1997): 1. Print.

"New Facility, Location Designed to Meet Long-Term Needs." *The SWITCH* (Feb.-Mar. 1993): 1. Print.

Alessandrini, Walter. "President's Perspective." *The SWITCH* (Jan.-Feb. 1995): 2. Print.

Alessandrini, Walter. "President's Corner." *The SWITCH* (Dec. 1990): 2. Print.

"Message from the President." *The SWITCH* (Nov.-Dec. 1998): 1. Print.

The SWITCH (May 1996): 3. Print.

"Division Returns to Its Former Name..." *WABCO HI-LITES* (May-June 1981). Print.

"Message from the President." *The SWITCH* (May-June 1998): 1. Print.

"Board Appoints President and Chairman." *The SWITCH* (Dec. 1990): 3. Print.

"News from the President." *The SWITCH* (Jan.-Feb. 1999): 1. Print.

"Our President Meets Our President." *The SWITCH* (Mar. 1991): 1. Print.

"News from the President." *The SWITCH* (July-Aug. 1999): 1. Print.

"One by One." *The SWITCH* (Apr.-May 1994): 1. Print.

Hayes, Ed. "Modest Buckeye Earned His Laurels." *Featured Articles From The Orlando Sentinel*. Web. 28 Feb. 2011.

"GEORGE W. BAUGHMAN, 90, 1225 Sara Court, Winter Park, Died..." *Orlando Sentinel* 11 Apr. 1990. *Orlando Sentinel Article Collections*. Web. 3 Mar. 2011.

"Was to Give Wood $5,000 For Vote on Big Contract." *The New York Times* 16 Dec. 1915: 1-6. Print.

Reuters. "Company News; American Standard." *The New York Times* 05 July 1988. *The New York Times*. Web. 4 Mar. 2011.

"More Men Needed Under 9-Hour Law." *The New York Times* 2 Mar. 1908. Print.

"New Strategy, Structure And Management Announced For Ansaldo Signal - Free Online Library." *Free Online Library*. 10 Mar. 1997. Web. 22 Mar. 2011. <http://www.thefreelibrary.com/New Strategy, Structure And Management Announced For Ansaldo Signal-a019190825>.

"Ticknor Named to Head American Standard's Worldwide Signaling Business." *PR Newswire* 13 Aug. 1987. Print.

Coleman, John Pressley. Railway Switching Apparatus. John Pressley Coleman, assignee. Patent 764,043. 5 July 1904. Print.

Prall, William E. Improvement In Pnuematic Signaling Apparatus. William E. Prall, assignee. Patent 175,750. 4 Apr. 1876. Print.

Robinson, William. Improvement in Electro-Magnetic Railroad-Signals. Patent 108,633. 25 Oct. 1870. Print.

Saxby, John. Patent 1479. 24 June 1856. Print.

Saxby, John. Patent 1479. 24 June 1856. Print.

ANSALDO SIGNAL NV. *Form 6-K Filed 1998-08-25*. *SEC Watch*. Web. 22 Mar. 2011.

Train Control Bulletin No. 1. Bulletin. American Railway Association, 1930. Print.

Bulletin 85. Bulletin. Union Switch & Signal. Print.

First Annual Report, Block Signal and Train Control Board (US). Rep. 1908. Print.

March 10, 1891 Minutes. Minute Book. Union Switch & Signal, 1891. Print.

Meeting Minutes August 1998. Minute Book. 1998. Print.

Bulletin No. 2. Bulletin. Union Switch & Signal. Print.

Bulletin No. 2. Bulletin. Union Switch & Signal. Print.

Exhibit K, D. of J. Statement. Statement. 1927. Print.

Patent Litigation List. Union Switch & Signal. Print.

January 1879 Minutes. Minute Book. Union Electric Signal, 1879. Print.

Bulletin 98. Bulletin. Union Switch & Signal. Print.

Statement. 1927. Print.

Transcript of Hearings. Transcript. 1922. Print.

Bulletin No. 26. Bulletin. Union Switch & Signal. Print.

Bulletin No. 92 Union Vane Relays. Bulletin. Swissvale: Union Switch & Signal, 1922. Print.

Bulletin 72. Bulletin. Union Switch & Signal. Print.

Train Control to D. of J. (Q.30 of 1926 Questionnaire). Statement. 1926. Print.

Bulletin No. 9. Bulletin. Union Switch & Signal. Print.

Rosenthal, Dave. *The Driverless Train Turns 50: The History of the 42nd Street Automatic Shuttle*. Tech. Atlanta: Metropolitan Atlanta Rapid Transit Authority, 2009. Print.

Bulletin 97. Bulletin. Union Switch & Signal. Print.

Bodnar, Stephen. *Introduction to Railroad Signaling: Railroad Systems Control and Operation*. Publication. Union Switch and Signal, 2006. Print.

July 28, 1883 Minutes. Minute Book. Union Switch & Signal, 1883. Print.

Report to Stockholders. Rep. 1885. Print.

First Annual Report, Block Signal and Train Control Board (US). Rep. 1908. Print.

Bulletin 80. Bulletin. Union Switch & Signal. Print.

March 10, 1891 Minutes. Minute Book. Union Switch & Signal, 1891. Print.

January 1879 Minutes. Minute Book. Union Electric Signal, 1879. Print.

Cooper, Michael G., and David M. Rosenthal. *What's Old Is New Again: MARTA's Improvement to Track Circuit Reliability and Maintenance for Mainline Interlockings*. Tech. Atlanta. Print.

Metropolitan Transportation Authority. *Award a Contract for P2550 Rail Parts*. Rep. Los Angeles: Metropolitan Transportation Authority, 2008. Print.

Train Control Bulletin #1. Bulletin. American Railway Association,

1930. Print.

Bulletin No. 9. Bulletin. Union Switch & Signal. Print.

Bulletin 96. Bulletin. Union Switch & Signal. Print.

March 8, 1892 Minutes. Minute Book. Union Switch & Signal, 1892. Print.

Statement. 1916. Print.

Lewis, L. V. *CTC*. Rep. 1944. Print.

White, C. C. Paper. Print.

Report to Stockholders. Rep. 1885. Print.

Meeting Minutes June 2000. Minute Book. 2000. Print.

Union Switch & Signal. *Minute Book No. 3*. Minute Book. Union Switch & Signal, 1951. Print.

Bulletin 89. Bulletin. Union Switch & Signal. Print.

Holtgren, Thor. *Railroad Travel in the State of Business*. Paper. New York: National Bureau of Economic Research, 1943. Print.

Farnsworth, Malcolm M. *The Union Switch and Signal Company: A Review of Its Predecessors, Formation, Developments, Growth, Activities, Acquisitions and Affiliates*. Rep. Union Switch and Signal, 1948. Print.

Meeting Minutes May 2001. Minute Book. 2001. Print.

Saxby & Farmer Interlocking. Catalogue. Union Switch & Signal, 1885. Print.

Bulletin No. 55. Bulletin. Union Switch & Signal. Print.

Bulletin 91. Bulletin. Union Switch & Signal. Print.

Bulletin 87. Bulletin. Union Switch & Signal. Print.

Saxby & Farmer Interlocking. Catalogue. Union Switch & Signal, 1885. Print.

Patent Agreement File 2077. File. Print.

July 28, 1883 Minutes. Minute Book. Union Switch & Signal, 1883. Print.

Meeting Minutes February 2001. Minute Book. 2001. Print.

March 8, 1892 Minutes. Minute Book. Union Switch & Signal, 1892. Print.

Patent Litigation List. Union Switch & Signal. Print.

"This Day in History - September 8, 1910." *The Has-Been*. 7 Sept. 2010. Web. 15 Feb. 2011.

Hill, Natalie. "People on the Move." *Post-Gazette.com*. 31 Aug. 2008. Web. 22 Mar. 2011.

"The Plant of the Union Switch & Signal Co., 1909." *The Has-Been*. 25 Dec. 2008. Web. 15 Feb. 2011.

Whitten, David O. "The Depression of 1893." *Economic History Services*. 01 Feb. 2010. Web. 11 Feb. 2011.

"Ansaldo STS - US&S Pens $30M China CBTC Contracts." *RedOrbit*. 18 Dec. 2008. Web. 22 Mar. 2011.

Slack, Brian. "Rail Deregulation in the United States." *Hofstra People*. Web. 10 Mar. 2011.

"The Harahan Memos." *The Has-Been*. 6 Nov. 2007. Web. 15 Feb. 2011.

Walters, John. "Bus Jacking The Revolution." *Questia*. Web. 03 Mar. 2011.

"Colonel Henry G. Prout, Has-Been." *The Has-Been*. 17 Mar. 2008. Web. 15 Feb. 2011.

"George Westinghouse Biography (1846-1914)." *How Products Are Made*. Web. 07 Feb. 2011.

Feurer, Rosemary. "Union Switch and Signal Strike of 1914." *Labor History Links*. Web. 16 Feb. 2011.

"Interesting Websites." *The Has-Been*. 02 Mar. 2011. Web. 18 Mar. 2011.

Woolley, John T., and Gerhard Peters. "Richard Nixon: Statement on Signing the Regional Rail Reorganization Act of 1973." *The American Presidency Project*. Web. 10 Mar. 2011.

"North Shore Connector." Web. 18 Mar. 2011.

OPRRMS. "Last "Armstrong" Interlocking in USA Retired 5/2/2010." *Altamont Press*. 05 May 2010. Web. 14 Feb. 2011.

Calvert, J. B. "The Block System and Its History." *Railways: History, Signalling, Engineering*. 19 Dec. 2005. Web. 14 Feb. 2011.

"Varlen Corporation — Company History." *FundingUniverse*. Web. 10 Mar. 2011.

"Political Interpretations of The Wonderful Wizard of Oz." *Wikipedia*. 18 Jan. 2011. Web. 11 Feb. 2011.

"Union Switch & Signal Strike Photograph Collection." *Historic Pittsburgh*. Web. 16 Feb. 2011.

"ATCS Evesdropping." *The Has-Been*. 19 Dec. 2010. Web. 18 Mar. 2011.

Baer, Christopher T. "PRR Chronology 1876." *The Pennsylvania Railroad Technical & Historical Society*. Apr. 2006. Web. 1 Feb. 2011.

"Railroad Interlocking Control System Having Shared Control of Bottleneck Areas - US Patent 5301906 Description." *PatentStorm: U.S. Patents*. Web. 18 Mar. 2011.

"The History of the Transformer." *The Edison Tech Center*. Web. 15 Feb. 2011.

"Thomas Mellon Evans." *Babson College*. Web. 11 Mar. 2011.

"The 1917 Swissvale Plant Fire." *The Has-Been*. 19 Jan. 2008. Web. 18 Feb. 2011.

"Railroad - The Robber Barons." *Law Library*. Web. 10 Mar. 2011.

"Railroad History, An Overview Of The Past." *The American Rail-roads*. Web. 10 Mar. 2011.

"About LeTourneau." *LeTourneau Technologies*. Web. 02 Mar. 2011.

"ConnDOT: Chapter 7 DOT History." *Connecticut Department of Transportation*. 9 Sept. 2003. Web. 03 Mar. 2011.

"Union Switch & Signal CEO Resigns." *The Business Journals*. 31 July 2007. Web. 22 Mar. 2011.

"GEFCO." *GEFCO - Mobile and Portable Drilling Rigs*. Web. 02 Mar. 2011.

"Gaulard and Gibbs Secondary Generator." *Museo Galileo*. Web. 15 Feb. 2011.

"Regional Rail Reorganization Act Law & Legal Definition." *US Legal*. Web. 10 Mar. 2011.

"History of South Station: Station History." *South Station*. Web. 10 Feb. 2011.

"CEO Puts Union Switch & Signal on New Track | Pittsburgh Business Times." *The Business Journals*. 4 Aug. 2008. Web. 22 Mar. 2011.

Covington, Edward J. "William Stanley, Jr." *Frognet.net*. Web. 7 Feb. 2011.

"Business: Steel: The Strike's Blow - TIME." *Time*. 26 Oct. 1959. Web. 07 Mar. 2011.

Monahan, John M. "Melpar." *World Lingo*. Ed. Martt Harding. Web. 2 Mar. 2011.

"The Lost and Found Dept." *The Has-Been*. 11 May 2008. Web. 15 Feb. 2011.

American Short Line and Regional Railroad Association. "Short Line Tax Credit Extension." *ASLRRA - American Short Line & Regional Railroad Association*. Web. 24 Mar. 2011.

Pinto, Jim. "JimPinto Weblog - Invensys." *JimPinto.com*. 8 Mar. 2011. Web. 18 Mar. 2011.

Bryan, Frank W., and Robert S. McGonigal. "Railroad Signals." *Trains Magazine*. 1 May 2006. Web. 15 Feb. 2011.

"Canaan, NH Train Wreck, Sept 1907" GenDisasters.com. 25 Jan. 2010. Web. 5 Apr. 2011.

"Locomotive engineers and firemen's monthly journal" Volume 1. Number 1. Feb. 1888.

The Editors of Publications International, Ltd. "Early Twentieth Century Railroads." *Howstuffworks*. Web. 11 Feb. 2011.

"John McGee: Executive Profile & Biography." *BusinessWeek*. Web. 22 Mar. 2011.

"Gaulard and Gibbs Secondary Generator." *Museo Galileo*. Web. 15 Feb. 2011.

Mark Mickelson & Company. "Railroad Track Maintenance Tax Credit - €"45G Tax Credit." *Mark Mickelson and Company*. Web. 24 Mar. 2011.

"Salmon O. Levinson." *The Has-Been*. 23 Oct. 2008. Web. 15 Feb. 2011.

"Boston - South Station, MA (BOS)." *Great American Stations*. Web. 10 Feb. 2011.

"Union Switch & Signal Garners Long Term Contract With CSX. - Free Online Library." *Free Online Library*. 14 Dec. 2000. Web. 22 Mar. 2011.

"Struble versus Young." *The Has-Been*. 20 Sept. 2009. Web. 18 Feb. 2011.

Bakke, Dave. "Union Pacific to Close Springfield's Ridgely Tower." *ArizonaRails.com*. 22 May 2010. Web. 14 Feb. 2011.

"Department of State Bulletin." *Welcome to the US Petabox*. Web. 28 Feb. 2011.

"Patent Number: US000183487." *United States Patent and Trademark Office*. Web. 1 Feb. 2011.

"Railroading Today." *The American Railroads*. Web. 10 Mar. 2011.

"Www.switch.com, 1996-2008." *The Has-Been*. 25 Oct. 2008. Web. 18 Mar. 2011.

Marano, Ray. "Manning the Switch." *Smart Business*. 29 July 2005. Web. 22 Mar. 2011.

"New Jersey Transit." *New Jersey Transit - Home*. 13 Feb. 2002. Web. 18 Mar. 2011.

"Ridgely Tower." *RR Signal Pix*. Web. 14 Feb. 2011.

"Walter D. Uptegraff, Has-Been." *The Has-Been*. 03 Feb. 2009. Web. 15 Feb. 2011.

Herron, Jim. "History Of The Semaphore." *E-Train*. Apr. 2002. Web. 14 Feb. 2011.

"Take Tea." *L'idea Magazine*. 2005. Web. 11 Feb. 2011. <http://www.lideamagazine.com/taketea.htm>.

Miller, Peter. "CSX "Almost There..." After a Long Train Ride." *Lexis Nexis*. 03 Mar. 2011. Web. 10 Mar. 2011.

"Eli J. Blake, Inventor of the Searchlight Signal." *The Has-Been*. 10 Jan. 2011. Web. 18 Mar. 2011.

"William E. Prall." *District Energy Library*. Jan. 1997. Web. 01 Feb. 2011.

"The Nation's Railroads - They Use Us, Then Abuse Us." *Brotherhood of Locomotive Engineers and Trainmen*. Web. 10 Mar. 2011.

Warner, John. "US& S Garners Another Contract for New Jersey Transit ASES Cab Signal Systems." *Swamp Fox*. 04 Jan. 2001. Web. 22 Mar. 2011.

"The Plant of the Union Switch & Signal Co., 1909." *The Has-Been*. 25 Dec. 2008. Web. 15 Feb. 2011.

"Facility Management Conference Program." Feb. 2008. Web. 22 Mar. 2011.

"Manhattan Project." *Spartacus Educational*. Web. 08 Mar. 2011.

"A Brief History of Le Roi." *Le Roi Tractair*. Web. 03 Mar. 2011.

"East of Eden." *The Has-Been*. 28 June 2009. Web. 15 Feb. 2011.

University of Guelph. Web. 7 Feb. 2011.

"Ansaldo STS USA Gets Port Authority Connector Contract." *Trib-LIVE*. 23 May 2009. Web. 18 Mar. 2011.

Longman, Phillip. "Back on Tracks." *New America Foundation*. Jan.-Feb. 2009. Web. 10 Mar. 2011.

"About Stevens: Colonel John Stevens III." *Stevens Institute of Technology*. Stevens Institute of Technology. Web. 26 Jan. 2011.

"Sentinels of the Rail." *Semaphores.com*. Mar. 1999. Web. 14 Feb. 2011.

"US&S Advertising Copy Written by Rudyard Kipling?" *The Has-Been*. 28 Feb. 2010. Web. 15 Feb. 2011.

"Brief Chronology of Railroad History, With Passenger Emphasis." *NARP: National Association of Railroad Passengers*. 1 Aug. 2006. Web. 10 Mar. 2011.

Toole, Randal O. "The Case for Privatizing Transit." *CATO Institute*. 18 Nov. 2010. Web. 3 Mar. 2011.

"Railroad Rehabilitation & Improvement Financing (RRIF)." *FRA*. Web. 23 Mar. 2011.

"Some Things Should Stay in Vegas." *The Has-Been*. 6 Dec. 2010. Web. 15 Feb. 2011.

"Project "REWIND"." *The Has-Been*. 19 July 2010. Web. 18 Mar. 2011.

"Wabtec Acquires G&B Specialties and Bach-Simpson." *The Has-Been*. 14 July 2010. Web. 18 Mar. 2011.

"ATCS Monitor for Windows." *ATCS Monitor for Windows*. 29 Nov. 2008. Web. 18 Mar. 2011.

"Herbert A. Wallace, Has-Been." *The Has-Been*. 25 Feb. 2008. Web. 15 Feb. 2011.

Margolin, Jed. "Road to the Transistor." *Jed Margolin's Web Site*. 1993. Web. 04 Mar. 2011.

"That Didn't Take Long...." *The Has-Been*. 10 Feb. 2011. Web. 18 Mar. 2011.

"George's Last Breath." *The Has-Been*. 4 July 2009. Web. 15 Feb. 2011.

"Labor History Sites to Visit While in Pittsburgh." *PA AFL-CIO*. Web. 16 Feb. 2011.

Butler, Chris. "FC111: The Start of the Industrial Revolution in Britain (c.1750-1800)." *The Flow of History*. 2007. Web. 26 Jan. 2011.

"On Time with Rail-Traffic Optimization Technology." *Advanced Technology Program Status Report Database*. Web. 18 Mar. 2011.

"Panic of 1893." *Historycentral.com*. Web. 11 Feb. 2011.

"New Leadership Announced at Union Switch & Signal and Ansaldo Signal. - PR Newswire | HighBeam Research: Online Press Releases." *HighBeam*. 20 Mar. 2001. Web. 22 Mar. 2011.